SMALL FIRMS IN REGION.
DEVELOPME:

Britain, Ireland and the United States

Over the last twenty years or so major changes have taken place in the attitude of governments towards small firms. No longer is the large firm regarded as the motor for growth and no longer is the emphasis upon scale economies at the plant level. Instead public policy is increasingly directed towards the smaller enterprise. Several European governments have, over the past five years, initiated for the first time programmes of assistance to small businesses, whilst the United States, which has a much longer history of encouraging small businesses, is planning an increased emphasis on the sector in its efforts to increase economic growth and employment.

Several groups are, however, starting to challenge this new orthodoxy and in this collection of papers the contribution of small businesses to economic development is assessed in a number of diverse localities.

The impact which policies designed to create employment and wealth in small firms can have, within a time scale such as ten years, is discussed by many of the authors, and it is clear that the impact of such policies is very long term. The evident divergencies between entrepreneurial behaviour in prosperous and less prosperous areas impact across a broad spectrum of businesses, and mitigate against the successful application of a set of narrow policy initiatives designed to encourage indigenous development. The small businessman is, by definition, independent, and faced with a massive diversity of problems. This, and the fact that the major contribution to economic development in a locality is made by a handful of firms, suggests that public policy cannot assist all small firms, and that it should be directed towards those firms whose improvement in performance (as related to public funds injected) has maximum benefit for the economy as a whole.

Dr David Storey is Senior Research Associate at the Centre for Urban and Regional Development Studies, the University of Newcastle upon Tyne.

SMALL FIRMS IN REGIONAL ECONOMIC DEVELOPMENT

*Britain, Ireland and the
United States*

Edited by

D. J. STOREY

University of Newcastle upon Tyne

for

REGIONAL STUDIES ASSOCIATION

The right of the
University of Cambridge
to print and sell
all manner of books
was granted by
Henry VIII in 1534.
The University has printed
and published continuously
since 1584.

CAMBRIDGE UNIVERSITY PRESS

Cambridge
London   New York   New Rochelle
Melbourne   Sydney

CAMBRIDGE UNIVERSITY PRESS
Cambridge, New York, Melbourne, Madrid, Cape Town, Singapore,
São Paulo, Delhi, Dubai, Tokyo

Cambridge University Press
The Edinburgh Building, Cambridge CB2 8RU, UK

Published in the United States of America by Cambridge University Press, New York

www.cambridge.org
Information on this title: www.cambridge.org/9780521125581

© Cambridge University Press 1985

First published 1985
Reprinted 1987
This digitally printed version 2009

A catalogue record for this publication is available from the British Library

Library of Congress Catalogue Card Number: 85–5773

ISBN 978-0-521-30198-5 Hardback
ISBN 978-0-521-12558-1 Paperback

Contents

List of Contributors   *page* vi

1 Introduction D. J. STOREY   1

2 Manufacturing employment change in Northern England
   1965–78: the role of small businesses D. J. STOREY   6

3 New firms and rural industrialization in East
   Anglia A. GOULD and D. KEEBLE   43

4 Spatial variations in new firm formation in the United
   Kingdom: comparative evidence from Merseyside, Greater
   Manchester and South Hampshire
   P. E. LLOYD and C. M. MASON   72

5 An industrial and spatial analysis of new firm
   formation in Ireland P. N. O'FARRELL and R. CROUCHLEY   101

6 Innovation and regional growth in small high technology
   firms: evidence from Britain and the USA R. P. OAKEY   135

7 Regional variations in capital structure of new small
   businesses: the Wisconsin case
   R. E. SHAFFER and G. C. PULVER   166

8 The world of small business: turbulence
   and survival A. R. MARKUSEN and M. B. TEITZ   193

9 The implications for policy D. J. STOREY   219

Index   231

Contributors

R. Crouchley *Dept of Sociology, University of Surrey*

A. Gould *Dept of Geography, Cambridge University*

D. Keeble *Dept of Geography, Cambridge University*

P. E. Lloyd *North West Industry Research Unit, University of Manchester*

A. R. Markusen *Institute of Urban and Regional Development, University of California*

C. M. Mason *Dept of Geography, University of Southampton*

R. P. Oakey *Centre for Urban and Regional Development Studies, University of Newcastle upon Tyne*

P. N. O'Farrell *Dept of Town Planning, University of Wales Insitute of Science and Technology*

G. C. Pulver *Dept of Agricultural Economics, University of Wisconsin–Madison*

R. E. Shaffer *Dept of Agricultural Economics, University of Wisconsin–Madison*

D. J. Storey *Centre for Urban and Regional Development Studies, University of Newcastle upon Tyne*

M. B. Teitz *Institute of Urban and Regional Development, University of California*

1

Introduction

D. J. STOREY

A number of different responses have been made by governments to the recession which has gripped the economies of many of the world's leading industrialised countries since the mid-1970s. Faced with both a declining total market and increased competition from LDCs most governments have offered financial and non-financial incentives for manufacturing industry to become more price competitive. Government policies have also stressed the importance of both moving into technologically more sophisticated products and of using more efficient production methods ('automate or liquidate'). The increased use of robots in manufacturing and the relatively slow rates of increase in labour productivity within the service sector have encouraged the belief that an increasing proportion of new employment will be created in services, based upon high incomes but low manning levels in manufacturing.[1]

Nowadays, at the forefront of any discussion of new employment initiatives and new productive developments is the small firm. Evidence presented by Birch (1979) and by Birch and MacCracken (1983) suggests that the large corporation cannot be regarded as the major source of new jobs, even though multi-national corporations as a group are significantly more important in influencing both the volume and nature of world trade than any single national government, or perhaps any collection of governments. Yet if large firms are viewed as playing a major role in economic development it is in conjunction with smaller enterprises.

Over the last twenty years major changes have taken place in the attitude of governments towards small firms. No longer is the large firm regarded as the motor for growth and no longer is the emphasis upon scale economies at the plant level. Instead public policy is increasingly directed towards the smaller enterprise. Several European governments have, over the past five years, initiated for the first time programmes of assistance to small business,[2] whilst the United States, which has a much longer history of encouraging small business, is planning an increased emphasis on the sector in its efforts to increase economic growth and employment.[3] Several groups, from different perspectives, are, however, starting to challenge this cosy consensus. One group has challenged the statistical and methodological basis of the view that small firms have been, and are likely to be in the future, a major source of new jobs (Armington and Odle (1982); Harris (1984); Fothergill and Gudgin (1979)). In particular the work of Birch and his colleagues has been challenged on the grounds that Birch has seriously overestimated the contribution of small business to employment creation. A second challenge comes from those who find public assistance to small business unacceptable on political grounds. Such individuals argue that since small firms provide low-paid, non-unionised jobs in poor working conditions they do not merit support. Indeed it is socially unacceptable for public monies or support to be provided to a group of capitalists which prides itself on virtues such as self-reliance and independence and whose avowed motives are to reduce public sector spending and business taxes (Rainnie (1985); GLC (1983)). Finally, and most relevantly in the context of these chapters, a third group has questioned the extent to which national policies have different impacts over space. For example it has been argued that the prime impact of small business policies is in prosperous, rather than less prosperous, areas (Storey (1982)) because of spatial differences in industrial structure, location of markets and differences in the socio-economic composition of the population.

In the following collection of papers the contribution of small businesses to economic development will be assessed in a number of localities. Three of the articles are concerned exclusively with the United Kingdom (Storey, Gould and Keeble, Lloyd and Mason), one with Ireland (O'Farrell and Crouchley), two with the United States (Shaffer and Pulver, Markusen and Teitz) whilst the article by Oakey compares locations in Britain and the USA. As well as national differences there are variations according to the level of economic prosperity of the areas investigated. For example Lloyd and Mason compare relatively prosperous South Hampshire with the far from prosperous Merseyside/Manchester area. Storey examines the impact of new firms on the economy of Northern England which has experienced amongst the highest unemployment levels in England for nearly half a century. In contrast Gould and Keeble quantify the contribution of new businesses in the '"amenity-rich", university-focussed areas of Southern England'. In the paper by Shaffer and Pulver the distinction between rural and urban areas in Wisconsin is stressed. There are also considerable variations in per capita income *within* Ireland.

In undertaking research on small business a variety of approaches can be used. It is possible, using statistical measures of aggregate economic activity such as output, employment, capital expenditure, etc., to identify the major trends in these variables according to the size of the enterprise.[4] This approach is favoured by economists interested in the competitive process, and the contribution to that process made by large enterprises. On the other hand Case Studies and interviews with firms are appropriate for highlighting some of the subtler developments and processes which can be 'lost' in aggregate studies.[5]

Each approach has its advantages and disadvantages. The aggregate approach can become mechanistic and is crucially dependent upon the availability of data suitable for the purpose, whereas interviews with firms can be misleading because the firms knowingly or unknowingly may mislead the interviewer, or the respondent may be ignorant of the factors in question. Furthermore some researchers are insufficiently careful to ensure that the sample of firms interviewed *are* representative of the population as a whole.[6]

The papers presented in this volume reflect these differences in approach. For example the papers by O'Farrell and Crouchley and Storey use aggregate data to examine the impact of new firms in Ireland and Northern England. These papers are essentially a statistical examination of the numbers of new firms, their sectoral and spatial distribution, with an attempt to link patterns to factors such as the size mix of establishments. Both papers also relate the employment created to the wider employment changes which were taking place in these economies.

The papers by Oakey, Markusen and Teitz, Shaffer and Pulver and Lloyd and Mason are based on interviews with small firms. All these authors point to the diversity of the small firm population and argue, implicitly or explicitly, that interviews are a particularly effective vehicle for obtaining views on the impact of issues such as financial constraints, labour availability, etc. The paper by Gould and Keeble perhaps straddles these two approaches since interviews were undertaken by the authors and these inevitably coloured the results presented, although the paper restricts itself primarily to aggregate analysis.

Given both the diverse methodological approaches and the different

4   D. J. STOREY

areas under study it is to be expected the contributions of small firms to local economic development would vary. Such variations could also be magnified by differences in the definitions used of a 'small' firm. For example, in the UK alone, Cross (1983) shows that government employs more than forty different definitions relating either to eligibility for assistance or the completion of statistical forms. At an international scale the definitional problems are particularly troublesome (Storey (1983)). It is for this reason that each contributor distinguishes carefully between the small firm and the new firm. Furthermore since the majority of studies in this volume are of manufacturing business the conventional definition of a small firm is an enterprise having less than 200 workers. Where the contributors use either a wider or a narrower definition this is identified in the text.

Whilst it is important to ensure broadly similar definitions of the term 'small firm', it would be wrong to impose definitions which arbitrarily exclude endemic sources of variation. If public policy towards small business, or towards groups of small firms, is to be formulated then a major pre-requisite is the recognition that policy will induce a variety of very different responses from this collection of 'strange bedfellows'. It is in the context of better informing that debate that the Regional Studies Association has agreed to reproduce in this volume the four papers which appeared in vol. 18 no. 3 of *Regional Studies* (Gould and Keeble, Lloyd and Mason, O'Farrell and Crouchley and Oakey), together with the three new papers by Storey, Markusen and Teitz and by Shaffer and Pulver.

NOTES

1 For a review of these trends see Gudgin (1983) or Robertson *et al.* (1982).
2 The most comprehensive review is contained in Economist Intelligence Unit (1983).
3 See Thompson and Leyden (1983).
4 Examples of this type of work are Utton (1979), Prais (1979), Hay and Morris (1984).
5 It should be noted that the approaches are not mutually exclusive. Hay and Morris, for example, conduct interviews with directors of small unquoted companies and conduct statistical analysis to compare the performance of quoted and unquoted companies.
6 The problems of defining 'new' firms for the purposes of survey analysis are clearly set out in Mason (1983). The problems of identifying a population from which to sample new firms are outlined in Smith and Storey (1985).

REFERENCES

Armington, C. and Odle, M. (1982) 'Small Businesses – How Many Jobs?', *Brookings Review*, Winter, pp. 14–17.

Birch, D. L. (1979) 'The Job Generation Process', MIT Program on Urban and Regional Change, Cambridge, Mass.

Birch, D. L. and MacCracken, S. J. (1983) 'Small Business Share of Job Generation: Lessons Learned from the Use of a Longitudinal File', MIT Program on Neighborhood and Regional Change, Cambridge, Mass.

Cross, M. (1983) 'The United Kingdom', in D.J. Storey (ed.), *The Small Firm: An International Survey*, Croom Helm, London.

Economist Intelligence Unit (1983) *The European Climate for Small Businesses: A Ten Country Study*, EIU, London.

Fothergill, S. and Gudgin, G. (1979) 'The Job Generation Process in Britain', Centre for Environmental Studies, Research Series no. 32.

Greater London Council (1983) 'Small Firms and the London Industrial Strategy', Economic Policy Group Strategy Document no. 4, London.

Gudgin, G. (1983) 'Job Generation in the Service Sector', Department of Applied Economics, University of Cambridge.

Harris, C. S. (1984) 'Small Business and Job Generation: A Changing Economy or Differing Methodologies', Harris and Associates, Washington DC.

Hay, D. and Morris, D.J. (1984) *Unquoted Companies*, Macmillan, London.

Mason, C. M. (1983), 'Some Definitional Problems in New Firms Research', *Area*, vol. 15, pp. 53–60.

Prais, S.J. (1979), *The Evolution of Giant Firms in Britain*, Cambridge University Press.

Rainnie, A. F. (1985), 'Small Firms, Big Problems: The Political Economy of Small Businesses', *Capital and Class*, no. 25, Spring, pp. 140–68.

Robertson, J.A.S., Briggs, J. M. and Goodchild, A. (1982) 'Structure and Employment Prospects of the Service Industries', *Research Paper no. 30*, Department of Employment, London.

Smith, E. and Storey, D. J. (1985) 'Defining a Population of New Firms', CURDS, University of Newcastle upon Tyne.

Storey, D. J. (1982) *Entrepreneurship and the New Firm*, Croom Helm, London.

Storey, D. J. (ed.) (1983) *The Small Firm: An International Survey*, Croom Helm, London.

Thompson, J. H. and Leyden, D. R. (1983), 'The United States of America' in D.J. Storey (ed.) *The Small Firm: An International Survey*, Croom Helm, London.

Utton, M. A. (1979), *Diversification and Competition*, Cambridge University Press.

2

Manufacturing employment change in Northern England 1965–78: the role of small businesses[1]

D. J. STOREY

I. INTRODUCTION

This paper examines the contribution to manufacturing employment change of small and medium sized enterprises (SMEs) in Northern England – an area which has experienced persistently high rates of unemployment for more than fifty years. The region has also been a recipient of almost every initiative introduced by national government to create employment, and whilst there have been periods of modest success the relative position of the Region has remained virtually unchanged during this half century.

In this paper a small business is defined as having less than 200 employees, although it is recognised that the problems and priorities for a well established business with 180 employees will differ markedly from those of the two-man business. Yet this heterogeneity is characteristic of the small firm sector and the main distinction to be made in these pages will be on the basis of age, i.e. between the new and the well established small business.

The first section of this paper documents the importance of SMEs in the Regions of the United Kingdom, but it then concentrates upon changes in employment in the three counties of Durham, Cleveland and Tyne & Wear which in 1978 contributed 78% of manufacturing jobs in Northern England. Having outlined the data sources to be used, the remainder of the paper is devoted to a detailed study of employment change, with particular emphasis upon the performance of SMEs.

2. SMALL BUSINESSES IN THE NORTHERN REGION

The Northern Region of England, defined as the administrative counties of Tyne & Wear, Cleveland, Durham, Northumberland and

Cumbria, has traditionally been regarded as an area at best un-sympathetic to the concepts of small scale enterprise and self-employment. The Northern Region Strategy Team (1977) regarded the low rate of new firm formation as a major explanation of the North's poor economic performance, illustrated by its persistently high rates of unemployment. The Region has been shown to have a smaller stock of small and medium sized enterprises (SMEs) than other areas of the UK, and a lower birth rate of SMEs. These points are illustrated in Table 2.1.

This table constructs, from a number of data sources, indices to illustrate Regional variations in both the existing stock of small firms and of changes in that stock. According to virtually all indices, the North of England not only has fewer small businesses than the remainder of the UK, but in 1980 had the lowest birth rate (except for Northern Ireland and Wales), the highest death rate (except for the South East of England) and the highest *net* reduction in stock of business.

Table 2.1 distinguishes manufacturing from all small enterprises. It shows that in the North 90.1% of manufacturing units employ less than 200 workers. This is the lowest of any UK Region except for Northern Ireland and markedly lower than the prosperous South East where 96.2% of manufacturing units employ less than 200 workers. Table 2.1 also normalises the number of manufacturing SMEs according to the number of males employed in manufacturing, since this group is most likely to form wholly new SMEs.[2] According to this index the North has 12.4 SMEs per thousand male manufacturing employees, sub-stantially less than any other UK Region including Northern Ireland. It is approximately half that of the UK as a whole, which is dominated by the South East where there are 29.5 SMEs per thousand male manufac-turing employees.

The emphasis upon manufacturing SMEs may lead to a biased picture since only about 10% of all SMEs are in the manufacturing sector. There are major variations between UK regions in the relative importance of manufacturing so the remainder of Table 2.1 examines data on all small employment units. It shows, in the final column, that the North had virtually the fewest SMEs per head of population of any Region in the UK in 1978, being approximately 20% below that of the South East Region.

These Regional differences would be unimportant if they were transitory. However, data on deaths and births, according to Value Added Tax (VAT) Registrations, suggests this is unlikely since, as

Table 2.1 *The stock, birth and death rates of SMEs by UK Region*

Standard region	Manufacturing units			All units[a]		
	Number of SMEs in 1978	SMEs as % of total manufacturing units 1978	SMEs per '000 male manufacturing employees 1978	Births of units liable for VAT 1980	Deaths of units liable for VAT 1980	Small units per '000 pop. 1978
Northern	3,845	90.1	12.4	+ 7.1	− 9.2	17.9
Yorkshire & Humberside	10,505	93.5	21.2	+ 8.0	− 7.8	19.2
East Midlands	8,527	93.4	21.8	+ 8.0	− 7.1	17.2
East Anglia	3,078	93.3	21.8	+ 8.8	− 8.0	18.8
South East	38,858	96.2	29.5	+ 9.7	− 9.9	21.4
South West	6,428	94.3	20.5	+ 7.5	− 7.4	21.3
West Midlands	12,427	93.3	17.4	+ 8.3	− 7.2	18.4
North West	14,207	93.6	20.2	+ 8.7	− 8.8	18.2
Wales	4,361	93.6	18.9	+ 6.8	− 6.7	18.2
Scotland	9,163	94.1	22.1	+ 7.2	− 7.3	19.1
Northern Ireland	1,348	90.0	14.8	+ 5.9	− 6.3	18.1[b]
Total	112,747					

[a] All units for 'Births' and 'Deaths' data include all groups except those not liable for payment of VAT, viz.: land, insurance, postal services, betting, gaming and lotteries, finance, education, health, burial and cremation, trade unions and professional bodies. For the column headed 'Small units per '000 population 1978' only agriculture and domestic services are excluded.

[b] No data are available from the Census of Employment Units. The estimate for Northern Ireland is based upon live VAT units in 1980.

Sources: manufacturing data are obtained from *Analysis of United Kingdom Manufacturing (Local) Units by Employment Size 1978*, BSO, Business Monitor PA 1003, 1981. Data on all units are obtained from Ganguly (1982b).

Data on population are obtained from *Regional Statistics 1980* and *Regional Trends 1982*, CSO, London.

noted earlier, Table 2.1 shows that in 1980 the North had virtually the highest death rate and virtually the lowest birth rate of any UK region. This form of Regional imbalance shows no sign of diminishing.

The Northern Region of England, according to these indices, has a lower than average stock of new businesses and appears to be showing few signs of reversing this trend. Although the indices generally point in the same direction, data on births and deaths of all establishments do not exist as a time series, so 1980 may or may not be typical of long term trends. There are also anomalies in using a register based upon VAT payments, partly because of the number of excluded sectors and partly because very small businesses do not pay VAT. It may also be difficult to distinguish changes of name from changes of business.[3] Nevertheless the uniformity of the results suggests the small business sector of Northern England is less developed than in most other UK Regions and shows little sign of catching up.[4]

3. THE AREA OF STUDY

The remainder of the paper will concentrate upon a proportion of the Northern Region. It will examine manufacturing employment change in the Counties of Cleveland, Durham and Tyne & Wear where, in 1978, 78% of manufacturing jobs in the Region were located. Table 2.2 shows the distribution of manufacturing employment in the five Counties which comprise the Region, sub-divided according to size of establishment. In the Region as a whole 41.3% of manufacturing employment, in 1978, was in establishments with more than 1,000 workers, whilst in Cleveland, Durham and Tyne & Wear the proportion was 41.9%

The Northern Region, and the three Counties under study, are therefore significantly less dependent upon small business than other Regions of the UK. The three Counties are more dependent upon metal manufacture, heavy engineering and shipbuilding than either the Region or the UK as a whole, so that contractions in the demand for labour by these industries will have a considerable impact upon total employment.

It is against this background that a review is made of the role of SMEs in creating employment during the period 1965–78. In the following two sections, the job accounting framework used is described together with the sources of data employed.

Table 2.2 *Total manufacturing employment in Counties of the Northern Region in 1978, by size of establishment*

Establishment size	Cleveland No.	%	Cumbria No.	%	Durham No.	%	Northumberland No.	%	Tyne & Wear No.	%	Total No.	Total Northern %
11–19	1,641	1.8	1,536	2.2	1,629	2.3	673	3.0	4,718	3.1	9,657	2.4
20–49	3,388	3.7	3,151	4.5	3,791	5.3	1,559	7.0	8,500	5.5	20,389	5.0
50–99	3,027	3.3	3,649	5.2	4,560	6.3	1,572	7.1	9,289	6.1	22,097	5.4
100–199	6,119	6.7	7,026	10.0	7,104	9.9	3,131	14.0	13,680	8.9	37,060	9.0
200–499	14,446	15.8	12,180	17.3	16,322	22.6	6,459	28.9	27,343	17.7	76,750	18.7
500–999	14,387	15.8	13,371	19.0	10,196	14.1	1,946	8.7	34,405	22.3	74,305	18.2
1000+	48,361	52.9	29,456	41.8	28,483	39.5	6,981	31.3	56,385	36.4	169,666	41.3
Total	91,369	100.0	70,369	100.0	72,085	100.0	22,321	100.0	154,320	100.0	409,924	100.0

Source: Business Monitor PA 1003, 1978.

4. JOB ACCOUNTS

Net employment change between two points in time may be attributable to several factors, viz. the closure of establishments, to new
employment being created either by the location of new branches in the
area, the expansion or contraction of existing firms or to the establishment of wholly new businesses.

The terms used throughout this chapter are shown in Figure 2.1.
Reading from the right it shows that *net job change*, which can be either
positive or negative, is an amalgam of influences. Most simply it
represents the net effect of the summation of *gross new jobs* created less
gross job losses. Figure 2.1 shows that gross new jobs may be sub-divided
between *openings* of new establishments and the *expansions* of existing
firms. An opening is defined to be an establishment which is new to any
one of the three Counties after 1965, whilst an expansion is defined as an
establishment which exists in both 1965 and 1978, and which employs
more workers in 1978 than in 1965. Gross job losses are defined to be
closures and *contractions*. Closures are establishments which existed in
1965 but which failed to survive until 1978, whilst contractions are
establishments which existed in 1965 and 1978, but which employed
less workers in 1978 than in 1965. Finally openings and closures are
sub-divided. Openings are sub-divided between births, i.e. wholly
new firms (the entrepreneur establishing his own business) and the
movement into one of the Counties of establishments which have
previously had a manufacturing presence elsewhere (*in-moves*).
Closures can be sub-divided between *deaths*, where an establishment
ceases to exist, and *out-moves* where operations are transferred to
another site outside the County in which it was located.

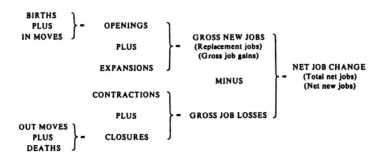

Fig. 2.1. The job generation process

Source: Centre for Environmental Studies paper, Policy Series 11.

The term *in-situ change* is also used. This is defined to be expansions less contractions, and refers to employment change in establishments which exist throughout the period in question. Armed with this terminology it can be seen that net job change is openings minus closures plus in-situ change.

5. MANUFACTURING EMPLOYMENT 1965–78: DATA SOURCES

An examination of employment change, in the form presented here, can only be undertaken when access is available to a data base on employment and other variables for each establishment. The data used herein are the property of the Metropolitan County Council of Tyne & Wear and of the County Councils of Durham and Cleveland.

For the three Counties the data base comprises virtually 5,000 establishments for each of which there are up to 120 units of information on employment, ownership, date of opening, closure, location, industry, etc. A complete statement of the data sources and format is available from the author.

To construct such a data base information has to be collected from a variety of sources and a number of problems arise in ensuring comparability over such a long period. Two important difficulties occur because of boundary changes and changes over the methods of collecting employment information. The County Councils in their current form have existed only since 1974 and their boundaries do not exactly coincide with those of the Employment Exchange Office Areas (EEOA) which collected employment data prior to 1971. The boundary overlaps between Counties and EEOAs can be seen from Figure 2.2.

Prior to 1971 the prime source of employment data used in the data base was the records of Principal Employers (PEs) which were held by the managers of each EEOA for establishments employing five or more people. Unfortunately these records do not exist for all EEOAs for all years but generally their coverage is good for the areas between the Rivers Tyne and Tees since this data was collected on a regular basis by the pre-1974 County Durham and then maintained by post-1974 Durham CC.[5] Data are also available for most years in Newcastle, but coverage for the post-1974 district of North Tyneside is weaker, and full coverage is not available until 1968. Data for earlier years has been provided from a variety of sources including the Factory Inspectorate, Northumberland CC and the Northern Region Strategy Team. For all years and for all districts prior to 1971, however, coverage of estab-

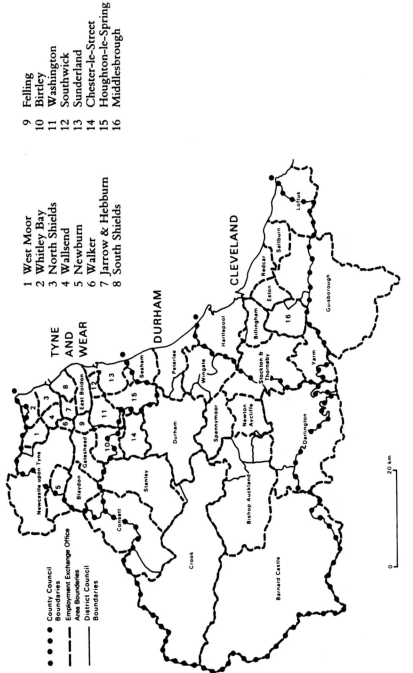

1 West Moor
2 Whitley Bay
3 North Shields
4 Wallsend
5 Newburn
6 Walker
7 Jarrow & Hebburn
8 South Shields

9 Felling
10 Birtley
11 Washington
12 Southwick
13 Sunderland
14 Chester-le-Street
15 Houghton-le-Spring
16 Middlesbrough

County Council
Boundaries
Employment Exchange Office
Area Boundaries
District Council
Boundaries

TYNE
AND
WEAR

DURHAM

CLEVELAND

Fig. 2.2 North-East England

lishments employing less than five workers is imperfect, but such limitations are common to virtually all UK Regional data bases.

To overcome the problem of imperfect coverage it was decided to identify years in which coverage was best and insert employment data where establishments were known to exist in manufacturing at the time, or for establishments which subsequently became manufacturers. The years chosen were 1965 and 1969 and where there were clear omissions employment data for the years nearest to 1965 and 1969 were inserted.[6]

For the Counties of Tyne & Wear and for Durham this procedure is satisfactory. For Cleveland, however, the main data sources for the year 1965 are the records of a comprehensive survey, *Teesside Survey and Plan*, together with data from the EEOA of Hartlepool and Hartlepool Headland. Coverage for the interim years until 1975 is patchy, after which comprehensive data again became available. This limitation is particularly relevant to data on establishments which open and close between 1966 and 1975.

Three other problems occur in presenting a satisfactory comparison of employment change over the period. The first was that the introduction of Selective Employment Tax meant a number of manufacturing establishments were re-classified in 1966/7 into the Services Sector. This is particularly important in the Newcastle area. Secondly changes in the 1958 Standard Industrial Classification (SIC) to the 1968 SIC also caused non comparability due to establishment re-classification.

The main difficulty, however, was the change in the basis of counting employment. From 1948 onwards employment was estimated on the basis of a quarterly count of exchanges of Class I National Insurance Cards, but the decision to phase these out in 1975 meant that a direct count of employment was needed. It was decided to mount a census which counted all individuals employed in June of each year. The census began in June 1971 and was, at that time, called the Annual Census of Employment (ACE). Non comparability between pre- and post-1971 data occurred where casual and seasonal workers were employed, or in the case of individuals holding more than one job.

Several further problems occur with ACE. The first is that coverage for 1971 and 1972 is incomplete because the records refer only to establishments which were in business in 1973, but since the PE records for Tyne & Wear and for Durham continued in 1971 and 1972 this does not present major problems. Secondly several establishments, classified as manufacturing according to the PEs, were re-classified as services in ACE. This may be due either to statistical re-classification or to real changes in the activities of the establishment. Unfortunately, it has not been possible to quantify the importance of these effects. A third problem is that even in 1978 there were establishments which, although in existence at the time of the census, provided no employment return. Such establishments cannot be regarded as closures, but the absence of employment data in 1978 makes it impossible to calculate in-situ change. Finally, in constructing the data base, no attempt is made to supplement the data provided in ACE after 1973 since this is supposed to be a comprehensive record of employment.

This section has stressed the difficulties and complexities of constructing a data base of this magnitude over a long period, but it should not be inferred that the data are significantly inferior to those produced from other Regional data bases which exist in the UK. Indeed its coverage of small establishments, the fact that it has annual data, its coverage of establishments, moving between the manufacturing and service sectors, and its ability to distinguish between enterprises and establishments distinguish it from other data bases currently available.

6. EMPLOYMENT CHANGE IN TYNE & WEAR, DURHAM AND CLEVELAND 1965–78

In the three Counties manufacturing employment declined by 50,877 between 1965 and 1978. This represents a fall of about 13.2% compared with a fall of 15.1% for the UK as a whole during this period. The performance of the three Counties varied considerably with Tyne & Wear, and Cleveland declining by 16% and 24% respectively, whilst Durham increased its employment by 14%.

Although County Durham increased its manufacturing employment, whilst Cleveland and Tyne & Wear contracted, Table 2.3 shows very clearly that gross job losses (contractions + closures) were very similar. In all three Counties gross job loss represented about 40% of the 1965 stock of jobs.

This is a most important finding since it offers statistical support for the work of Birch (1979) who argued that across Regions in the United States rates of gross job loss were similar. According to Birch the characteristic of successful Regions was the numbers of new jobs created. Hence policy to create employment should ignore job losses, which were 'inevitable', and concentrate upon the creation of new employment. Before accepting such a conclusion, however, it is necessary to undertake further disaggregation of employment change.

In the remainder of this section employment changes in the three Counties taken together will be examined according to industrial sector,

Table 2.3 *Manufacturing job accounts, 1965–78*
Durham, Cleveland and Tyne & Wear

	Manufacturing employment in 1965	Openings	Closures	Expansions	Contractions	Gross new jobs	Gross job losses	Net change
Durham	66,373	+19,459	−12,673	+15,404	−12,629	+34,863	−25,302	+9,561
(% of 1965 employment)		(+29.3%)	(−19.1%)	(+23.2%)	(−19.0%)	(+52.5%)	(−38.1%)	(+14.4%)
Cleveland	114,485	+14,225	−14,574	+5,084	−32,301	+19,309	−46,875	−27,566
(% of 1965 employment)		(+12.4%)	(−12.7%)	(+4.4%)	(−28.2%)	(+16.9%)	(−40.9%)	(−24.19%)
Tyne & Wear	203,905	+23,329	−31,508	+21,564	−46,257	+44,893	−77,765	−32,872
(% of 1965 employment)		(+11.4%)	(−15.4%)	(+10.6%)	(−22.7%)	(+22.0%)	(−38.1%)	(−16.1%)

Notes: Openings: employment in 1978 in establishments not in the County in 1965.
Closures: employment in 1965 in establishments not in the County in 1978.
Expansions: the increase in employment in 1978 in establishments in the County in 1965 and in 1978, and which increased employment between those dates.
Contractions: the decrease in employment in 1978 in establishments in the County in 1965 and in 1978, and which decreased employment between those dates.

size of establishment and enterprise type. Categorising the discussion in this way should not suggest that these groups are entirely independent. For example it is likely that large establishments are more likely to be part of a group of companies than smaller establishments. Nevertheless, given the only limited degree of overlap, an examination of employment change, according to each dimension, does offer useful insights into the processes at work.

a. Employment change by industry sector

Table 2.4 shows an industrial categorisation for establishments which are *exclusively* in the manufacturing sector.[7] It shows that, for this group, net job loss was 51,317 during the period 1965–78. During that time, 56,311 jobs were created in openings of new establishments and 58,488 jobs were lost in closures. Amongst establishments which existed both in 1965 and 1978, 41,935 jobs were created in expansions, but 91,075 were lost in contractions.

Table 2.4 shows that of the net 51,317 jobs lost in the area, 18,807 or 37% were net losses in metal manufacturing, whilst a further 8,509 were lost in mechanical engineering and 5,757 in shipbuilding. The metal and metal-using industrial groups therefore accounted for 64% of the net job losses (or 41.1% of gross job losses) in the Region during this period.

Table 2.4 demonstrates, however, that net job losses occurred for very different reasons in shipbuilding, metal manufacture and mechanical engineering. In broad terms, although job losses through closure of mechanical engineering establishments represent 17.3% of all closures, and are the highest of any industry, these are virtually replaced by new openings which provide 9,961 new jobs. Similarly the industry provides 21% of all new jobs in expansions and 18.8% of jobs lost in contractions. Mechanical engineering is, therefore, an active sector with a rapid turnover of jobs, but one experiencing a modest net decline.

This contrasts with the metal manufacturing and shipbuilding sectors. These two sectors provide only 6.7% of new jobs in openings compared with 20.5% of job losses in closures. Amongst in-situ establishments the prime source of job loss is in metal manufacturing, where net losses are 14,090 jobs, rather than in shipbuilding where net losses were less than 2,300.

The sectors showing a positive employment growth during these years were instrument engineering, metal goods n.e.s., vehicles, paper and printing and other manufacturing. Despite this welcome growth

Table 2.4 Manufacturing job accounts, 1965–78, by industrial order: total employment
Tyne & Wear, Durham and Cleveland

Order	Description	Closures	%	Openings	%	In-situ	Expansions	%	Contractions	%	Gross new jobs	%	Gross job loss	%	Net job change
III	Food, drink & tobacco	-7,235	12.4	+5,180	9.2	-1,349	+2,450	5.8	-3,799	4.2	+7,630	7.8	-11,034	7.4	-3,404
IV	Coal & petroleum	-2,223	3.8	+3,699	6.6	-9,534	+2,801	6.7	-12,335	13.5	+6,500	6.6	-14,558	9.7	-8,058
V	Chemical & allied	-7,895	13.5	+3,178	5.6	-14,090	+1,069	2.5	-15,159	16.6	+4,247	4.3	-23,054	15.4	-18,807
VI	Metal manufacture															
VII	Mechanical engineering	-10,144	17.3	+9,961	17.7	-8,326	+8,792	21.0	-17,118	18.8	+18,753	19.1	-27,262	18.2	-8,509
VIII	Instrument engineering	-132	0.2	+2,188	3.9	-90	+103	0.2	-193	0.2	+2,291	2.3	-325	0.2	+1,966
IX	Electrical engineering	-1,784	3.1	+9,785	17.5	-9,581	+5,126	12.2	-14,707	16.1	+14,911	15.2	-16,491	11.1	-1,580
X	Shipbuilding	-4,084	7.0	+616	1.1	-2,289	+4,909	11.7	-7,198	7.9	+5,525	5.6	-11,282	7.5	-5,757
XI	Vehicles	-2,080	3.6	+1,201	2.1	+1,775	+2,659	6.3	-884	1.0	+3,860	3.9	-2,964	2.0	+896
XII	Metal goods n.e.s.	-2,087	3.6	+3,769	6.7	-195	+1,788	4.3	-1,983	2.2	+5,557	5.7	-4,070	2.7	+1,487
XIII	Textiles	-3,356	5.7	+4,243	7.5	-3,903	+1,302	3.1	-5,205	5.7	+5,545	5.7	-8,561	5.7	-3,016
XIV	Leather															
XV	Clothing & footwear	-7,342	12.5	+4,751	8.4	-2,625	+1,805	4.3	-4,430	4.9	+6,556	6.7	-11,772	7.9	-5,216
XVI	Bricks, pottery glass, cement	-3,328	5.7	+898	1.6	+403	+2,173	5.2	-1,770	1.9	+3,071	3.1	-5,098	3.4	-2,027
XVII	Timber & furniture	-3,762	6.4	+1,811	3.2	-1,012	+958	2.3	-1,970	2.2	+2,769	2.8	-5,732	3.8	-2,963
XVII	Paper & printing	-1,987	3.4	+1,813	3.2	+1,942	+5,109	12.2	-3,167	3.5	+6,922	7.0	-5,154	3.5	+1,768
XIX	Other manufacturing	-1,049	1.8	+3,218	5.7	-266	+891	2.1	-1,157	1.3	+4,109	4.2	-2,206	1.5	+1,903
	Total	-58,488	100.0	+56,311	100.0	49,140	+41,935	100.0	-91,075	100.0	+98,246	100.0	-149,563	100.0	-51,317

none of these industries made a major impact upon replacing jobs lost in the steel, steel-using and chemical industries.

Perhaps the most promising sign, from the viewpoint of the Regional economy as a whole, is that the electrical engineering industry is a major source of new jobs in openings, providing 9,785 or 17.5% of all jobs in manufacturing – only slightly less than that created in mechanical engineering. In-situ establishments, however, contracted in employment with a net loss of more than 9,500 jobs.

b. Employment change by size of establishment

Employment change can also be examined according to size (in terms of numbers of workers) of establishments. These data are provided in Table 2.5 which includes all establishments, i.e. including those which for at least part of their existence were in the services sector.

The final column shows that net job change for the Counties was 50,877 but, most interestingly, the figure above shows that the area incurred a net job loss of 51,141 in establishments employing more than 1,000 workers.

The final column of Table 2.5 clearly illustrates the major changes which took place in the establishment size structure of the area with, on balance, increased employment in all the smaller sizes of establishments. On the other hand, the larger establishments show a decline of employment with the major source of job loss being in those employing 1,000 or more workers in 1965. Indeed if this group of establishments were excluded manufacturing employment in the area would actually have increased.

Table 2.5 also illustrates that the vast majority of job losses in giant establishments are due to contractions rather than closures. Giant establishments lost 9,181 jobs in closures, compared with 59,512 in contractions. Conversely job losses in closures were significantly more likely to occur in establishments employing between 100 and 500 workers.

Finally it should also be clear that although there are net job gains in small establishments these are swamped by job losses in the larger establishments. Thus the net increase in employment in establishments employing less than 100 workers was 11,878 but this was virtually eliminated by the job losses in establishments employing between 500 and 999 workers, and was only just over one fifth of the net job loss in giant establishments.

Table 2.5 *Manufacturing job accounts, 1965–78; by size of establishment: total employment Tyne & Wear, Durham and Cleveland*

Size Group	Closures	%	Openings	%	In-situ	Expansions	%	Contractions	%	Gross new jobs	%	Gross job loss	%	Net job change
1–9	−1,409	2.4	+2,820	4.9	+1,621	+1,977	4.7	−356	0.4	+4,797	4.8	−1,765	1.2	+3,032
10–24	−3,522	6.0	+5,421	9.5	+2,609	+3,648	8.7	−1,039	1.1	+9,069	9.2	−4,561	3.0	+4,508
25–49	−5,236	8.9	+5,812	10.2	+2,678	+3,988	9.5	−1,310	1.4	+9,800	9.9	−6,546	4.4	+3,254
50–99	−6,161	10.5	+7,224	12.7	+21	+2,975	7.1	−2,954	3.2	10,199	10.3	−9,115	6.1	+1,084
100–249	−12,097	20.6	+11,020	19.3	+2,785	+9,084	21.6	−6,299	6.9	+20,104	20.3	−18,396	12.3	+1,708
250–499	−12,151	20.7	+12,286	21.6	−1,855	+6,033	14.3	−7,888	8.7	+18,319	18.5	−20,039	13.4	−1,720
500–999	−8,998	15.3	+5,199	9.1	−7,803	+4,026	9.6	−11,829	13.0	+9,225	9.3	−20,827	13.9	−11,602
1,000+	−9,181	15.6	+7,231	12.7	−49,191	+10,321	24.5	−59,512	65.3	+17,552	17.7	−68,693	45.7	−51,141
Total	−58,755	100.0	+57,013	100.0	−49,135	+42,052	100.0	−91,187	100.0	+99,065	100.0	−149,942	100.0	−50,877

Note: this table includes establishments which moved between the manufacturing *and* the service sector.

c. Employment change by enterprise type

Tables 2.4 and 2.5 presented data according to the size and industry of establishments, but a small establishment may be owned by a large enterprise. Hence in Table 2.6 an attempt is made to distinguish employment change according to type of enterprise. The table makes the broad distinction between legally independent enterprises and those which are subsidiaries. Within each broad grouping it distinguishes the single plant enterprise which, by definition, is locally controlled, from the non-locally controlled. In this context 'local' is defined to be within the County, but in the vast majority of cases non-local refers to ownership located in South East England or abroad. Finally a distinction is made between locally controlled branches and locally controlled headquarters plants.

Table 2.6 shows that although small establishments showed on average an increase in employment, single plant independent firms, as a group, shed more than 2,500 jobs over the period. Locally based enterprises (i.e. INDSPF + INDLCHQ + INDLCBR) in fact perform poorly in losing a net total of more than 37,000 jobs.

Much of this job loss in INDLCHQ and INDLCBR is due to the presence within the group of the metal manufacturing sector which in 1965, prior to its Nationalisation, was in locally based private companies. However the net loss of more than 2,500 jobs in single plant independent businesses suggests that the true small *business* sector performed less well than the performance of small *establishments* in Table 2.5 might have suggested (Armington and Odle (1982)).

The prime source of new jobs in the Region can clearly be seen from Table 2.6 to be externally owned subsidiary companies. This is the classic 'branch plant' attracted to the Region by the availability of labour, suitable sites and government financial assistance, particularly in the late 1960s when the UK economy was relatively buoyant. It is clear from the table that it is the openings of such plants, rather than the performance of plants in existence in 1965, that is the major source of new jobs. Non-locally controlled subsidiary plants and non-locally controlled independent plants in fact provided virtually 60% of all new jobs in openings.

7. WHOLLY NEW MANUFACTURING FIRMS IN TYNE & WEAR, DURHAM AND CLEVELAND

The prime focus of any study of small firms is the wholly new firm. The characteristics of those founding new firms in Northern England have

Table 2.6 *Manufacturing job accounts, 1965–78; by establishment type: total employment Tyne & Wear, Durham and Cleveland*

Firm type	Closures	%	Openings	%	In-situ	Expansions	%	Contractions	%	Gross new jobs	%	Gross job loss	%	Net job change
INDSPF	−13,930	23.7	+10,802	18.9	+537	+8,792	20.9	−8,255	9.1	+19,594	19.8	−22,185	14.8	−2,591
INDLCHQ	−2,916	5.0	+1,711	3.0	−16,782	+2,786	6.7	−19,568	21.5	+4,497	4.5	−22,484	15.0	−17,987
INDLCBR	−8,957	15.2	+1,958	3.4	−9,978	+1,800	4.3	−11,778	12.9	+3,758	3.8	−20,735	13.8	−16,977
INDNLC	−8,351	14.2	+8,024	14.1	−7,357	+4,932	11.7	−12,289	13.5	+12,956	13.1	−20,640	13.8	−7,684
SUBSPF	−9,045	15.4	+6,264	11.0	−9,817	+4,510	10.7	−14,327.	15.7	+10,774	10.9	−23,372	15.6	−12,598
SUBLCHQ	−3,047	5.2	+376	0.7	−901	+792	1.9	−1,693	1.9	+1,168	1.2	−4,740	3.2	−3,572
SUBLCBR	−1,372	2.3	+1,648	2.9	−632	+558	1.3	−1,190	1.3	+2,206	2.2	−2,562	1.7	−356
SUBNLC	−9,024	15.4	+25,817	45.3	−5,655	+15,940	37.9	−21,595	23.7	+41,757	42.2	−30,619	20.4	+11,138
NAT/OTHER	−2,113	3.6	+413	0.7	+1,450	+1,942	4.6	−492	0.5	+2,355	2.4	−2,605	1.7	−250
Total	−58,755	100.0	+57,013	100.0	−49,135	+42,052	100.0	−91,187	100.0	+99,065	100.0	−149,942	100.0	−50,877

INDSPF   Independent Single Plant Firm.
INDLCHQ   Independent Locally Controlled Headquarters Plant.
INDLCBR   Independent Locally Controlled Branch.
INDNLC   Independent Non Locally Controlled Branch.
SUBSPF   Subsidiary, Single Plant Firm.
SUBLCHQ   Subsidiary, Locally Controlled Headquarters Plant.
SUBLCBR   Subsidiary, Locally Controlled Branch.
SUBNLC   Subsidiary, Non Locally Controlled.
NAT/OTHER Nationalised or Other Establishment.

been extensively reported by Storey (1982) so this section will only examine the contribution of such enterprises to employment in the three Counties.

The analysis is undertaken for 1,145 wholly new manufacturing firms created in the three Counties. The firms included have primarily local directors, partners or are sole proprietorships and were established for the first time in the three Counties after 1965. It should be stressed that since coverage of firms which ceased trading before 1976 is weak in Cleveland, it is likely that there will have been approximately 1,200 wholly new manufacturing firms created in the three Counties over thirteen years. Those wholly new firms which had a very short life, i.e. less than twelve months, are also under-represented, but in all other respects it is a census rather than a sample.

Of the 1,145 new manufacturing firms known to have been formed over a thirteen year period between 1965 and 1978, 887 or 77% survived until 1978, and it was possible to obtain employment data in 1978 for 774 of them. As noted above, in addition to under-enumeration in Cleveland, some new small firms survive for an insufficiently long period to appear in government statistical records. The failure rates given here may therefore underestimate the 'true' failure rate. It will also be shown that whilst a high proportion of new businesses survived until 1978 this was because many were less than five years old.

The industrial distribution of new businesses is shown in Table 2.7. The importance of new firms in the metal trades is clear, with 220 new firms being found in mechanical engineering and 153 in the miscellaneous metal industries. Other industries which generated large numbers of new firms were timber and furniture and paper and printing, with these, together with the metal trades, accounting for 57% of all new firms.

The rate at which wholly new firms are added to the stock of local independent firms is shown in column 2. Here an index is formulated by taking the number of wholly new firms founded since 1965 and dividing these by the number of locally owned independent businesses in the industry in 1965. Whilst this ignores closures and ownership change, it identifies industries where the formation of new firms was fastest, i.e. where additions to the stock of single plant independent firms was most rapid. According to this criterion, rates were highest in electrical engineering, clothing and footwear and shipbuilding, and lowest in food, drink and tobacco, metal manufacture, bricks and pottery and in paper and printing.

The employment created by wholly new firms is a major factor in the

Table 2.7 Openings of wholly new independent manufacturing firms, 1965–78, by industry
North-East England

Order	Description	Number of wholly new firms	New firm formation rate	No. of surviving firms — All	No. of surviving firms — With employment data	Employment in 1978 — Males FT	Males PT	Females FT	Females PT	Total	Average employment in survivors	% of all jobs in openings by industry	Failure rate
III	Food, drink & tobacco	46	0.21	30	28	149	10	82	48	289	10.3	7.4	0.34
IV	Coal & petroleum	44	1.26	31	27	255	5	143	67	470	17.4	12.7	0.29
V	Chemical & allied	20	0.36	16	14	191	5	8	6	210	15.0	6.6	0.20
VI, VII	Metal manufacture / Mechanical engineering	220	1.12	175	163	2,514	51	211	93	2,869	17.6	28.9	0.20
VIII	Instrument engineering	32	1.28	29	28	378	4	35	26	443	15.8	20.2	0.09
IX	Electrical engineering	74	1.80	57	50	557	4	231	46	838	16.8	8.6	0.23
X	Shipbuilding	23	1.48	15	12	136	17	11	9	173	14.4	28.1	0.35
XI	Vehicles	24	0.71	23	17	168	8	18	15	209	12.3	17.4	0.04
XII	Metal goods n.e.s.	153	1.21	125	114	1,196	11	99	45	1,351	11.9	35.8	0.18
XIII, XIV	Textiles / Leather	35	0.66	26	22	130	5	92	37	264	12.0	6.2	0.26
XV	Clothing & footwear	88	1.95	66	57	169	6	958	284	1,417	24.9	29.8	0.25
XVI	Bricks, pottery etc.	42	0.45	28	25	280	0	31	19	330	13.2	36.7	0.33
XVII	Timber & furniture	157	0.69	117	97	842	15	230	48	1,135	11.7	62.7	0.25
XVIII	Paper & printing	116	0.57	94	80	502	21	261	114	898	11.2	49.5	0.19
XIX	Other manufacturing	54	0.98	39	31	513	10	141	43	707	22.8	22.0	0.28
	Non-manufacturing	17	–	16	9	144	0	93	17	254	28.2	–	0.06
	Total	1,145		887	774	8,124	172	2,644	917	11,857	15.3	21.0	0.23

Note: FT = Full time.
PT = Part time.

current interest in small businesses. Table 2.7 shows that in the period 1965–78, new firms, by 1978, had created 11,857 jobs, of which 8,124 were full time male jobs. This was an arithmetic mean of 15.3 employees per surviving firm and new firms provided 21% of new jobs in openings. By 1978 for every 100 manufacturing workers in the Region less than four were employed in a wholly new manufacturing firm created since 1965.

The arithmetic mean size in 1978 of a wholly new manufacturing firm founded after 1965 varies from more than 20 workers in clothing and footwear (mainly female jobs) and in 'other' manufacturing to only eleven workers in timber and furniture and in paper and printing.

There are also major variations between industries in the importance of wholly new firms in job creation. Recalling that the Region is a major recipient, during this period, of employment in branch plants the relative importance of new firms in fact reflects the presence or absence of new jobs in branch plants rather than the importance of new firms per se.

Nevertheless it is clear that wholly new firms provided about one fifth (21%) of all new jobs in openings in the three Counties. In only two industries did new firms provide more than 40% of all new jobs – in timber and furniture and in paper and printing – yet these were the industries in which arithmetic mean employment size of the wholly new firm was smallest!

Finally the failure rates, by industry, are outlined in the final column. It shows that for all new firms, slightly less than one quarter (23%) of firms formed after 1965 did not survive until 1978. It also shows that there are again significant differences between industries. Failure rates are highest in shipbuilding (35%), food, drink and tobacco (34%) and bricks and pottery (33%). The lowest failure rates were found in instrument engineering (9%) and vehicles (4%). In most other industries the failure rates were close to the overall average.

The arithmetic mean employment in 1978 for wholly new firms created after 1965 in the three Counties was fifteen employees. The distribution of employment creation is, however, skewed to the right, as is illustrated by Table 2.8. This shows that of the 774 wholly new firms which provided employment data in 1978, 429 or 55.4% employed less than ten workers. Conversely 47 or 6% of all surviving new firms provided 33.8% of jobs in wholly new firms. The table also illustrates the low probability of a wholly new firm formed after 1965 having reached 100 employees by 1978. In the three Counties there are a total of eight such enterprises out of a known total of 1,145 births. As

Table 2.8 *Employment in 1978 in surviving openings of wholly new manufacturing firms Tyne & Wear, Durham and Cleveland*

Employment size in 1978	Number of firms	% of survivors	Employment in 1978					% of total 1978 employment in new firms in each size group
			Males		Females		Total	
			FT	PT	FT	PT		
1–9	429	55.4	1,296	55	289	222	1,862	15.7
10–24	217	28.1	2,324	49	645	279	3,297	27.8
25–49	81	10.5	1,865	57	568	203	2,693	22.7
50–99	39	5.0	1,477	11	942	199	2,629	22.2
100+	8	1.0	1,162	–	200	14	1,376	11.6
Total	774		8,124	172	2,644	917	11,857	100.0

Note: FT = Full time.
PT = Part time.

noted earlier, because of under-representation in Cleveland the total number was likely to have been approximately 1,200, so that the chances of reaching 100 employees over the period was approximately 0.6%. This is almost identical to that derived from other areas of the UK where comparable data are available (Storey 1982).

8. BIRTHS OF WHOLLY NEW MANUFACTURING FIRMS SINCE 1965 IN DURHAM AND TYNE & WEAR

One explanation for the low levels of employment created in wholly new firms established after 1965 could have been that many had not reached maturity by 1978. Indeed the rate of new firm formation has risen appreciably in recent years as can be seen from a study of Figures 2.3 and 2.4 which show the number of wholly new firms created in Durham and Tyne & Wear respectively in each year since 1965. Whilst there are some differences between Figures 2.3 and 2.4, both clearly

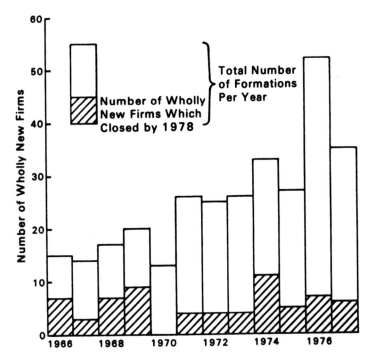

Fig. 2.3 Formation of wholly new manufacturing firms in County Durham, 1966–77

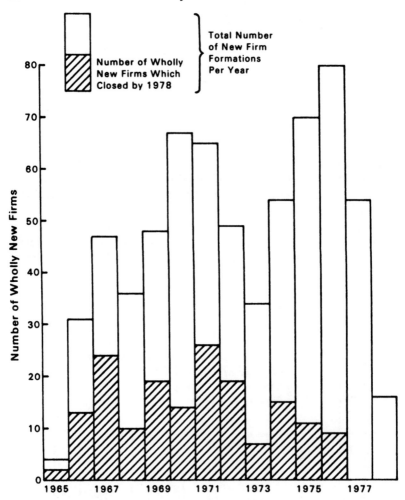

Fig. 2.4 Formation of wholly new manufacturing firms in Tyne & Wear, 1965–78

demonstrate that wholly new firm formations have fairly consistently increased throughout the period, with 1976 being a peak year for formations in both Durham and Tyne & Wear.[8]

The patterns of new firm formation vary only slightly between the two areas. In Tyne & Wear both 1970 and 1971 are years with a high rate of new firm formation, but in both areas more than one third of all new firms founded between 1965 and 1978 were established in the four years 1974–7 inclusive.

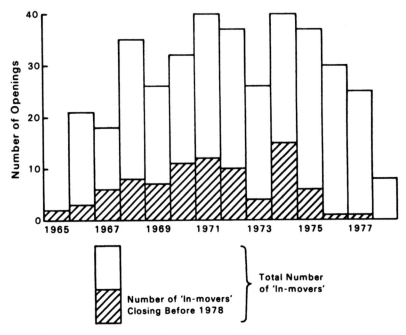

Fig. 2.5 Openings of manufacturing establishments in Tyne & Wear, excluding wholly new firms, 1965–78

The hatched areas of Figures 2.3 and 2.4 show the number of openings of wholly new firms which failed to survive until 1978. Thus of the fifteen openings of wholly new firms in County Durham in 1966 just over half failed to survive until 1978, whereas in Tyne & Wear twelve out of thirty-one openings failed to survive.

Time series data on openings of wholly new firms shows a marked bunching in the 1974–7 period, and contrasts sharply with an otherwise comparable time series of openings of all establishments, excluding wholly new firms. These are shown in Figures 2.5 and 2.6 for Tyne & Wear and Durham respectively. They show that for Durham the peak years for openings are 1969, 1971 and 1974, with a majority of openings in the 1968–72 period. In Tyne & Wear the years 1968–72 have a large number of openings although both 1974 and 1975 are also years when establishment openings are high.

A major explanation for this difference in pattern is that openings of wholly new firms are inversely related to the level of economic activity, whereas 'other' openings are positively related. During the 1960s and

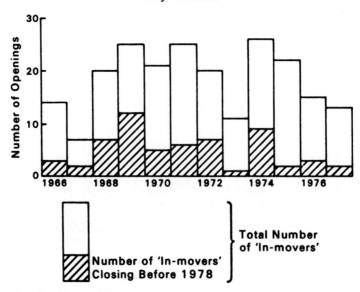

Fig. 2.6 Openings of manufacturing establishments in County Durham, excluding wholly new firms, 1966–77

early 1970s in conditions of prosperity and when the package of financial assistance to companies was attractive, there were a large number of enterprises wishing to expand into the North to take advantage of the financial incentives available. The oil crisis of 1973 and the subsequent slowing of economic growth, together with the associated increase in unemployment, meant that less firms were expanding. The Region was no longer able to attract in-moving companies on the same scale and individuals unable to obtain work with such companies found their only alternative to unemployment was to found their own company.[9] The decline in the number of in-movers is therefore the mirror image of the increase in the rate of new firm formation.

9. THE GROWTH AND DEATH OF NEW MANUFACTURING FIRMS IN DURHAM AND TYNE & WEAR

It will be recalled from Table 2.7 that the arithmetic mean employment in 1978 of wholly new firms formed in Tyne & Wear and Durham since 1965 was approximately 15.3 workers. This, however, may somewhat underestimate the average firm size since, as noted from Figures 2.3 and 2.4, new firm formation rates were highest at the end of the period. A

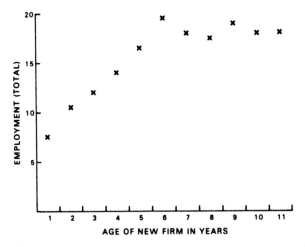

Fig. 2.7 Average employment in new firms by age in County Durham

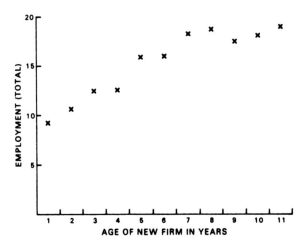

Fig. 2.8 Average employment in new firms by age in Tyne and Wear

proportion of new firms, therefore, will have had insufficient time to reach their ultimate employment size.

The employment growth of new manufacturing firms is traced and plotted in Figures 2.7 and 2.8 for Durham and for Tyne & Wear respectively. These two figures show a strikingly similar pattern. Figure 2.7 for Durham shows that in the year after formation arithmetic mean employment was seven workers and that employment rose fairly continuously until its sixth year at which peak employment was achieved, with nineteen workers. After that point from year six to year eleven there was, on balance, no further increase in employment. A similar pattern is observable for wholly new firms in Tyne & Wear except that growth in early years is slightly less rapid and the peak employment is not achieved until the eighth year. Nevertheless in both areas employment in firms eleven years old is no higher than in firms seven years old.

The general pattern, therefore, is for firms to have virtually half their 'ultimate' (year eleven) employment within the first year and to have two-thirds of their ultimate employment in the third year. This is explained by firms having to reach a minimum size quickly in order to reap available scale economies, but the achievement of these within six years has significant implications for those in the public sector whose task is to stimulate employment in small firms. These results make it clear that the main creators of gross new jobs, in the small firm sector, are new firms less than six years old. Once firms are more than six years old, *as a group*, they are unlikely to add significantly to gross new job creation at least for the following five years.

Whilst new firms which are more than six years old, on average, show no tendency to increase employment, it is clear from Table 2.8 that a handful of new firms continue to grow rapidly throughout the period. It is therefore important to distinguish between the perform- ance of the mean and median wholly new firm. Table 2.9 provides data for median employment levels of wholly new firms of different ages. A distinction is also made between new firms founded between 1965 and 1971 and those founded after that date since data coverage for the earlier period is weaker for the very small firms. To provide a satisfactory number of data points, new firms in Tyne & Wear and in Durham are combined.

Table 2.9 shows that the median level of employment in new firms is similar to the mean in the first year but the median firm then shows very little growth in employment after that time. After year four median employment is approximately two-thirds that of the mean and there appears to be no tendency for median employment to increase after the *fourth* year for new firms founded prior to 1971. For those founded after 1971 there does appear to be some growth but the data for years six and seven should be treated with caution since they are derived from relatively few data points.

A further explanation of the absence of employment growth in the median firm could be that surviving and non-surviving new firms have

Table 2.9 *Median employment in wholly new firms, by age*
Tyne & Wear and Durham

Age of firm (years)	Number of employees in new firm founded 1965–71	Number of employees in new firm founded after 1971
1	7.5	6.5
2	9.0	5.5
3	9.0	6.0
4	10.0	10.0
5	10.0	10.0
6	10.0	7.5
7	11.0	10.5
8	11.5	
9	11.0	
10	11.0	

been combined. To disaggregate these effects Figures 2.9 and 2.10 show for Durham and Tyne & Wear respectively the cumulative probability of a new firm ceasing to trade by a given year. The data for Durham show an interesting pattern with approximately half of all new firms founded in a given year having ceased within fourteen years of start-up. This is likely to be an underestimation since firms which exist only for a matter of weeks or months are unlikely to be recorded in the data set. The probability of death increases with age fairly continuously until year six with about 30% of new firms having ceased to trade by that time. There is then a marked 'flattening-out' of the curve between years six and eleven, after which it begins to rise again so that by year fourteen approximately half of the new firms will have ceased.

It will be recalled from the examination of employment change over time that peak employment was achieved in year six and the observation that the death rates also flatten out at that point, albeit temporarily, suggests that the two may be linked. For example, death rates are likely to be high in early years as the weak new firms are weeded out quickly, but as employment grows and the firm becomes better established it is much less likely to go out of business. In many respects the major hurdle of establishment has been overcome and there may be a tendency for the founder to concern himself less with growth, at least in terms of taking on new workers, and more with stability. He may feel that the taxation system offers insufficient incentive to grow and may be tempted to devote less time to his business. It appears, however, that

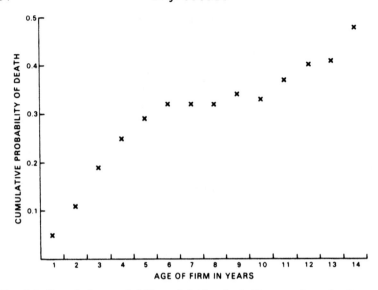

Fig. 2.9 Cumulative probability of death of wholly new firms in County Durham

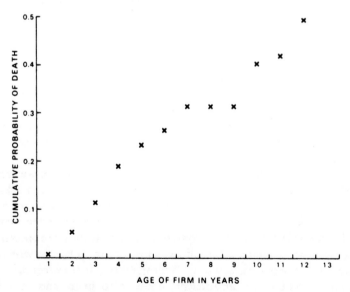

Fig. 2.10 Cumulative probability of death of wholly new firms in Tyne & Wear

whilst a strategy of non-growth may be maintained for a few years, the probability of death increases again from years nine to ten onwards. These failures could be explained by the founder having increasingly lost touch with the business once it had overcome its difficulties in the first six years of existence.

Clearly other explanations for these trends are possible. For example changes of ownership by take-over may occur for new businesses which have existed for five years and have an attractive track record. This may affect their subsequent probability of growth or death. It is also possible that the results are determined by groups of firms founded in specific years, so more testing of these hypotheses is necessary. Nevertheless the importance of distinguishing between surviving and non-surviving businesses is clearly important.

When data on probability of death and on employment growth in wholly new firms are compared they provide crude support for the view that wholly new firms undergo a life cycle of change. Enterprises start in order to test the market and many inevitably fail in early life. In fact a higher proportion fail than are shown here since their lifespan is too short to enter government statistical records. In several respects the fifth and sixth years of a new business appear to be something of a watershed. The new firm will have grown rapidly during this period and, having reached that stage, also appears to be relatively secure for a few more years. At least 30% of those businesses started at the same time will have failed but, in many respects, this is also a time of danger. The fifth and sixth years are not only the period when the chances of ceasing to trade fall, but also the years in which employment growth slackens. For the median firm it may even have ceased by year four.

It is possible to rationalise these findings by suggesting that during the first five or six years the new firm struggles to become established, but, once that objective is achieved, momentum disappears. Growth in terms of employment, and probably output, so much a feature of earlier years, is perhaps sacrificed for other objectives such as the golf course or a higher income (or both) and although the business continues to exist this sows the seeds for its subsequent demise.

Since failure, particularly in early life, is so characteristic of new firms it is important to be able to distinguish whether there are differences in the employment profiles of surviving as opposed to failed companies. For the purposes of definition 'failure' refers to a new firm which ceases to trade within six years, whereas a survivor is one that survives beyond that date.

The data on employment growth is shown in Table 2.10. This shows

Table 2.10 Employment in new firms in given years
Tyne & Wear and Durham

| | Firm started 1965–71 | | | | Firm started 1972–8 | | | |
| | Means | | Medians | | Means | | Medians | |
	Failures	Survivors	Failures	Survivors	Failures	Survivors	Failures	Survivors
Year 1	13.9	7.9	9.5	8.5	7.4	12.3	6.0	5.5
Year 2	16.8	11.7	10.0	8.0	8.4	11.7	6.0	5.0
Year 3	17.6	16.1	11.0	9.5	8.5	11.6	7.0	6.0
Year 4	10.6	16.1	8.0	10.5	8.7	14.1	7.0	8.0
Year 5	10.8	16.8	8.0	12.0	8.6	18.4	9.0	10.0
Year 6	8.06	17.2	8.0	11.0	1.6	21.1	–	9.5

that a new firm, started at any stage between 1965 and 1972 but which failed within six years, has a mean employment of 13.9 in its first year and a median of 9.5 workers. For new firms started in the pre-1972 period, in terms of mean and median values, the failures seemed to start with more employees than the survivors and continue to employ more workers, at least until their third year.

New firms founded after 1971 may reflect more accurately the growth patterns of the current population of new firms. Table 2.10 shows that for all years the mean employment amongst post 1971-established survivors exceeds the mean employment for failures. Again, however, this is attributable to the survival of a few rapid growth firms, since an examination of median employment shows little difference between survivors and failures.

IO. EPILOGUE

Detailed analysis of employment change in the Region cannot be undertaken after 1978 since this is the most recent data which are currently available. It would, however, be inappropriate to end at this point without some recognition of the magnitude of the changes which have taken place in the economy of Northern England since that date.

For the purposes of this research a detailed examination was made of establishments existing in 1978 to determine whether or not they had closed by 1982. The scale of the job loss found is illustrated in Table 2.11 which shows the job accounts for each four year period for Tyne & Wear and for Durham, since 1965.

The table shows that since 1978, 20.5% of manufacturing employment in Durham has been lost in closures alone with this representing 15,385 jobs. It represents a rate of job loss in closures more than two and a half times that for any other comparable period. For Tyne & Wear there is also a progressive increase in job losses through closures for each successive four year period since the mid-1960s, with the job losses since 1978 likely to be at least 50% higher than for any other comparable period.

It is not possible to present data for each period for Cleveland, but since 1978 the County has lost 15% of its manufacturing employment in closures. In total, therefore, in the post-1978 period, closures alone have resulted in a loss of more than 43,000 jobs, and it places in some context the 11,800 jobs created in wholly new firms in *thirteen* years, or the net increase of perhaps 12,000 jobs in establishments employing less than 100 workers over that period.

Table 2.11 *Manufacturing job accounts*
Tyne & Wear and Durham

	Openings		Closures		Expansions		Contractions	
	Durham	Tyne & Wear	Durham	Tyne & Wear	Durham	Tyne & Wear	Durham	Tyne & Wear
1965–9	+7.3	+3.5	−7.9	−2.9	+17.7	+10.7	−6.6	−12.3
1969–74	+9.7	+4.6	−7.4	−6.4	+20.0	+9.7	−11.4	−14.8
1974–8	+5.9	+4.0	−6.6	−5.9	+6.7	+8.1	−14.3	−16.6
1978–82	?	?	−20.5	−8.0[a]	?	?	?	?

Note: The table shows % changes on base year employment.
[a] The figure for Tyne & Wear is a minimum estimate.

II. CONCLUSION

Between 1965 and 1978 the three Counties of Cleveland, Durham and Tyne & Wear experienced a net loss of more than 50,000 manufacturing jobs, or 13% of employment. However this paper demonstrates that employment did not decline uniformly in all sectors, Counties or sizes of establishment. For example manufacturing employment in Durham increased by 14% over the period whilst employment in Cleveland and Tyne & Wear fell by 24% and 16% respectively. The major uniformity of the employment performance of the Counties was that, by 1978, *each* had lost approximately 40% of their 1965 employment through contractions and closures. An examination of the employment performance of the different sized *establishments* shows that a net loss of more than 50,000 jobs was incurred by giant establishments employing more than 1,000 workers whereas smaller establishments, in aggregate, showed an increase in employment over the period. It should not be assumed that these smaller establishments were necessarily locally owned single plant independent firms,[10] since an examination of employment change according to enterprise type showed the locally owned sector performed poorly in terms of employment creation. The paper makes it clear that the major source of new jobs in the three Counties was the non locally owned subsidiary (the branch plant). This type of establishment, attracted to the Region by a combination of both high levels of aggregate demand and Regional policy incentives, made a net contribution of more than 11,000 jobs over the period.

In many respects, however, matters changed both over the period under study and perhaps even more sharply since 1978. The world recession triggered, at least in part, by increased energy prices, accelerating inflation and a concern, at least by British governments, to reduce public spending, have contributed to a fall in aggregate demand within the UK and a reduction in the value of financial incentives to potentially mobile plants. This has coincided with (or perhaps caused) a shift in emphasis towards smaller scale, more locally based enterprises which are now seen to be a major source of the significantly fewer new jobs now being created in UK depressed Regions.

Of particular interest is the role of wholly new firms in creating employment. It is shown that 1,145 such enterprises created, between 1965 and 1978, a total of 11,857 jobs by 1978. This means that such enterprises provided work for only 4 in every 100 workers in 1978. Alternatively the 11,857 jobs in openings of wholly new firms may be set against the *net* decline of 18,807 jobs in metal manufacturing alone. It should be clear that whilst wholly new firms in all sectors of manufacturing were a worthwhile source of new jobs their formation rates have to rise 59% just to offset *net* job losses in metal manufacturing alone.

The data also illustrate the diversity of growth patterns of wholly new firms. It is shown that the majority of wholly new firms formed since 1965, surviving in 1978, had less than ten employees at that time. Conversely substantial job creation was found in relatively few new firms, with 33.8% of jobs in wholly new firms being created by only 6.0% of survivors. The chances of an individual starting a business which has 100 employees within a decade is between a half and three-quarters of 1%.

An examination of employment growth and death also illustrates this diversity of performance. It shows firstly that most wholly new firms grow quickly to their ultimate employment size with two-thirds of ultimate employment being achieved by the third year. Secondly it is found that at least 6% of wholly new firms starting in any one year die within twelve months and that at least 50% die within fourteen years. Thirdly, wholly new firms, which are more than six or seven years old are as likely to contract as expand, in contrast with the generally consistent expansion of young firms. Finally, when the employment growth performance of failures and survivors is compared, the failures generally show a much flatter growth pattern than the survivors.

For those concerned exclusively with public policy or employment and industrial policy the main inferences are:

It would be unwise, on the basis of past trends, to believe that in the Northern Region of England the small manufacturing sector can create sufficient employment within an acceptable time span to more than partially offset the job losses in large plants in the Region.

Northern England is culturally and economically less suited to taking advantage of initiatives to assist SMEs than any other UK Region.

Although SMEs' contribution to new employment in the region is always likely to be modest, assistance to SMEs may have to be pursued in the absence of other employment initiatives.[11] This paper has demonstrated the importance of *targeting* assistance to small and new firms, and the futility of an across-the-board policy to assist small firms.

It is of paramount importance to know the characteristics of the relatively few small firms likely to create large numbers of jobs, and avoid failure. The beginnings of this important work are outlined above and include the recognition that young firms are most likely to create jobs quickly. It is also shown that low growth firms are more likely to fail than high growth firms, whilst failure rates are significantly higher in, for example, food, drink and tobacco than in instrumental engineering. More work on targeting needs to be undertaken.

Finally the warning must be issued that whilst small firms policies, such as encouraging more individuals to start in business, may appear to create new jobs, there is always a risk of existing jobs being 'displaced' in other local small firms. The *net* effect of initiatives is therefore the relevant criterion for evaluating the success of SME policies.

NOTES

1 Earlier versions of this work on each of the Counties are available in Cleveland County Council (1982), Durham County Council (1982), Tyne & Wear County Council (1982). This work was undertaken in conjunction with these local authorities, in accordance with monitoring of their structure plans, under the Town & Country Planning Act of 1971. The permission of Eurostat to reproduce work undertaken for them is also gratefully acknowledged, but opinions expressed are those of the author alone. In undertaking this research invaluable assistance on data analysis was provided by Peter Hanson, Susan Howstan, Mark Tallintire, Robert Watson and Andrew Jones. Financial assistance was provided by the Gatsby Charitable Foundation, Cleveland County Council, Lloyds Bank, Imperial Chemical Industries and Eurostat.
2 Fothergill and Gudgin (1984) discuss this matter in some detail.
3 For a detailed description of the limitations of the data base see Ganguly (1982a).

4 A comprehensive analysis of the Regional aspects of small firm development in the UK is given in Storey (1982).
5 In 1974 local government in Britain was re-organised and a number of
major boundary changes took place. The pre–1974 County of Durham
stretched from the south bank of the River Tyne to Stockton in the south. In
the south of the Region there existed the County Borough of Teesside
which broadly consisted of the EEOAs of Middlesbrough and Stockton.
6 The years 1965 and 1969 are also chosen because they encompass an 'upturn'
and a 'downturn' in the trade cycle.
7 Data in Table 2.4 *exclude* establishments which are, for only part of their
existence, in the manufacturing sector.
8 The increase in wholly new firm formations reflects changes at a national
level. For example the annual average number of new companies incorporated in the 1966–70 period was 26,065. In the 1971–5 period it was
47,704 and in the 1976–9 period it was 57,656. This data is given in Storey
(1982).
9 In studies of new firms founded between 1973 and 1978 in Cleveland the
author found that 26% of founders claim to have been unemployed
immediately prior to starting their business – Storey (1982). It should be
recognised that during this period even Regional rates of unemployment
were only between 5% and 8%.
10 The distinction between enterprises and establishments is also made by
Armington and Odle (1982).
11 The author's views on an appropriate employment strategy for Northern
England are set out in Storey (1983). In particular there is evidence of an
opportunity for job losses to be reduced through management buy-outs. A
comprehensive evaluation of this initiative can be found in Coyne and
Wright (1984).

REFERENCES

Armington, C. and Odle, M. (1982) 'Small Business – How many Jobs?',
Brookings Review, Winter, pp. 14–17.
Birch, D. L. (1979) 'The Job Generation Process', MIT Program on Neigh-
borhood and Regional Change, Cambridge, Mass.
Cleveland County Council (1982) 'Manufacturing Employment Change in
Cleveland since 1965', Planning Department, Cleveland County Council,
Middlesbrough, Cleveland.
Coyne, J. and Wright, M. (1984) *Management Buy-Outs*, Croom Helm,
London.
Durham County Council (1982) 'Manufacturing Employment Change in
Durham since 1965', Planning Department, Durham County Council,
Durham.
Fothergill, S. and Gudgin, G. (1984) 'Geographical Variation in the Rate of
Formation of New Manufacturing Firms', *Regional Studies*, vol. 18, no. 3.
Ganguly, P. (1982a) 'Births and Deaths in Firms in the UK in 1980', *British
Business*, 29 January, pp. 204–7.

Ganguly, P. (1982b) 'Regional distribution of Births and Deaths in the UK in 1980', *British Business*, April 1982, pp. 648–50.

Northern Region Strategy Team (1977) *Strategic Plan for the Northern Region: Economic Development Policies*, HMSO, Newcastle.

Storey, D. J. (1982) *Entrepreneurship and the New Firm*, Croom Helm, London.

Storey, D. J. (1983) 'Regional Policy in a Recession', *National Westminster Bank Review*, November, pp. 39–47.

Tyne and Wear County Council (1982) 'Manufacturing Employment Change in Tyne and Wear since 1965', Planning Department, Tyne and Wear County Council, Newcastle upon Tyne.

3

New firms and rural industrialization in East Anglia

ANDREW GOULD and DAVID KEEBLE

INTRODUCTION

New manufacturing firms have in the last few years become an increasingly important focus of academic debate and government policy, in Britain as in other advanced capitalist industrial countries. Indeed, in terms of job generation and through their postulated role in fostering healthy and diverse local economies, notably through the introduction of new technology, they have been viewed by some commentators as a key to national economic recovery in the long run. In the context of the current recession recent published work suggests an increase in the number of new company registrations in Britain (Binks and Coyne, 1983; Ganguly, 1982c) which has been optimistically interpreted by some as indicating a revival of free enterprise entrepreneurship and the first step towards economic regeneration in the 1980s. Moreover, studies of such areas as the East Midlands (Fothergill and Gudgin, 1982) and Northern England (Storey, 1983) have found that employment in *small* establishments, which include many new firms, grew during the 1970s notwithstanding high small firm closure rates, whereas large plants recorded heavy job losses. Country-wide evidence on this point is however less clear-cut (Macey, 1982).

Until recently the relatively few British studies which have examined spatial variations in new manufacturing firm formation have dealt predominantly with the role of conurbations and large cities in generating new enterprise (Firn and Swales, 1978; Howick and Key, 1979; Lloyd, 1980; Lloyd and Dicken, 1979, 1982; London Industry and Employment Group, 1979; Nicholson and Brinkley, 1979). These studies indicate that there are wide differences in firm formation rates both between and within conurbations. It had been suggested by studies of Clydeside (Cameron, 1973) and Greater London (Wood, 1974) that the inner city areas of these conurbations appeared to be playing a declining role as generative environments for new manufacturing firms. Lloyd and Dicken (1979) have argued, however, that inner city areas are still an important source of manufacturing firms, with the inner area of Greater Manchester providing 29% of all conurbation jobs in new local manufacturing enterprises between 1966–75 and the inner area of Merseyside providing over 50% of that conurbation's new firm employment over the same period.

Only limited research has been carried out on the extent and nature of new manufacturing firm foundation in rural areas. Gudgin's pioneering work on the components of manufacturing employment change in the East Midlands between 1947–67 (Gudgin, 1978) and its subsequent update to 1975 (Fothergill and Gudgin, 1979, 1982; Gudgin et al., 1979) has established that firm formation rates are significantly higher in the rural areas of the East Midlands than in the region's cities and large towns. In addition to this urban–rural contrast, however, there is a marked negative relationship between the rate of firm formation and the proportion of an area's employees working in large factories. Mason's work in South Hampshire has revealed a similar urban–rural contrast (Mason, 1982). Firm formation rates between 1971–9 are shown to have been lowest in the cities of Portsmouth and Southampton and highest in the rural areas, notably the residentially attractive area of the New Forest. The apparently greater importance of new manufacturing firm formation in rural areas of Scotland has also been recently documented by Cross (1980, 1981), whose establishment data-bank reveals that new firms were the most important source of job generation in Scotland's twelve most rural employment office areas.

This paper presents the first results of an SSRC-funded project which examines the nature and extent of new manufacturing firm formation in the predominantly rural region of East Anglia between 1971–81. Since the 1960s, East Anglia has been the fastest growing industrial region in the UK with manufacturing employment increasing between 1965 and 1974 by 23% (from 167,000 to 205,000) before declining with recession to 182,000 in 1981 (*Employment Gazette*, various dates). Evidence is presented on the scale and structure of new manufacturing firm formation which indicates that in a national context, East Anglia appears to provide an exceptionally conducive environment for enterprise creation. In addition, the paper investigates the temporal pattern of new firm formation during the period, the nature of and reasons for marked spatial variations and rural bias in formation rates, and the distinctive nature and locational clustering of new 'high-technology' firms.

DATA AND METHODOLOGY

The definition and identification of new manufacturing firms has always posed well recognized difficulties for researchers (Mason, 1983). The definition adopted here, in line with most other work, is based on a conception of the new firm as 'one which has no obvious parent in any existing business enterprise' (Allen, 1961, p. 28). Such a definition makes clear the distinction between subsidiaries set up by existing companies and new independent firms. The latter are the sole concern of this paper. Independence has been interpreted in a purely legal sense, although in some cases there will undoubtedly be functional dependence in factor and product markets. All firms which were established as subsidiaries or established independently and subsequently acquired have been excluded from this study. The date of formation of the new independent firms studied is measured as being when the firm first registered as a business.

The validity and significance of the research presented here is rooted in arguably one of the most comprehensive and accurate databanks of surviving new firms so far developed in Britain. This databank records each legally independent new manufacturing firm which was created in East Anglia (the three counties of Cambridgeshire, Norfolk and Suffolk) between 1 July 1971 and 30 June 1981, provided that it was still operating at the latter date. In common with most other new firm studies, the databank thus refers to *surviving* new firms, and records their 1981 employment, Minimum List Heading classification (1968 SIC) and product description, and original East Anglian location if the firm had moved within the region after formation.

The databank was developed from lists of every manufacturing establishment present in East Anglia in 1971 and 1981, which include details of their employment, industrial classification and legal corporate status. These lists were compiled from extensive research into all available official and commercial sources, notably the comprehensive Factories Inspectorate establishment records for 1971 held by the Centre of East Anglian Studies of the University of East Anglia, current unpublished central government establishment records for the East Anglian counties, 1981 Market Location Directories, the Cambridgeshire County Council Directory of Manufacturing Industry, and Chambers of Commerce and CoSIRA (Council for Small Industries in Rural Areas) records. The procedure for identifying new firms involved an initial comparison of the two lists of firms which revealed those new to East Anglia between 1971–81. This list of potential new firms was then checked with 1971 Telephone Directories to eliminate those which were originally present, with Department of Industry Records of Openings and Closures in East Anglia to eliminate in-migrant firms, and with Dun and Bradstreet's *Who Owns Whom* to eliminate new companies which were not legally independent.

This detailed and laborious checking procedure yielded a list of over 2,100 apparently new manufacturing firms. However, in a procedure which has been adopted, to the best of the present authors' knowledge, in only one other new firm study (Mason, 1982), each of these firms was then telephoned by the project team to verify directly its date of formation, East Anglian origin, and status both as a manufacturing firm and as an independent company. This telephone survey also provided a more detailed product description and an accurate and up-to-date 1981 employment value. Although relatively expensive and time-consuming the importance of this major additional verification procedure in the identification of new firms is indicated by the very large number of apparent new firms eliminated. Fully 1,400 firms, or two-thirds, were removed from the original list, because they proved not to be manufacturing firms (300 firms), not new within the time period (350 firms), not independent but subsidiaries or branches of other firms (200 firms), not indigenous but in-migrant (100 firms), or simply not traceable despite an additional British Telecom check and hence no longer trading (500 firms). Interestingly, quite a number of those eliminated on the second of these grounds proved to be *too* new for inclusion, in the sense that they were established more recently than 30 June 1981, the terminal date for the study. This supports the view that the data sources used for compiling the initial 1981 lists were reasonably up-to-date.

The final East Anglian new firm databank thus represents one of the most carefully checked, authoritative and up-to-date data sets of surviving new manufacturing firms in existence in Britain. Equally, the surprisingly high rate of elimination consequent upon the telephone survey does inevitably suggest the possibility of some inflation of previously-calculated firm formation rates for other areas where these have been based solely on official and commercial records.

NEW FIRM FORMATION: SCALE, STRUCTURE AND COMPARISONS

The survey procedure identified a total of 703 manufacturing firms as having been started independently in East Anglia in the period July 1971 to June 1981 and surviving as independent companies to 1982 when the telephone check was carried out (Table 3.1). Thus, in 1981 new firms comprised 18% of the estimated 1981 population of manufacturing establishments in East Anglia, but provided only 4.7% of the region's total manufacturing employment, or 8,478 jobs. As with all previous new firms studies, therefore, and notwithstanding the smaller scale and greater vitality of manufacturing activity in the region compared with most other parts of Britain, the present research confirms that new firms have had only a very small impact on job generation in the region over the last decade. In the short run, and taking East Anglia as a whole, new manufacturing firms appear to have provided very few jobs relative to the rest of the region's manufacturing – and service – industry.

This said, the largest single concentration of new firms occurs in Cambridgeshire with 313 (45% of the total) surviving new firms having been formed there (Table 3.1). In order to take account of the size of the existing industrial base, firm formation is expressed as a rate – the number of surviving new firms formed in an area per 1,000 manufacturing employees in the base year of 1971. It is apparent that the highest firm formation rates also occur in Cambridgeshire, with East Cambridgeshire experiencing the remarkably high rate of 21.3 new firms per one thousand manufacturing employees in 1971. High rates are also recorded in Breckland, North Norfolk, Forest Heath and Suffolk Coastal.

In terms of the industrial composition of new East Anglian manufacturing firms, it is clear from Fig. 3.1 and Table 3.2 that the largest numbers are to be found in the mechanical engineering industry (order 7), with considerable totals also recorded in paper, printing and publishing (order 18), other metal goods (order 12), timber and furniture (order 17), electrical engineering (order 9) and other manufacturing industries (order 19). The last two of these are, however, more significant for employment than their numbers alone suggest, being third and fourth after mechanical engineering and printing in this respect. The above-average size of electrical engineering and other manufacturing firms indicated by this may also suggest a greater growth propensity for new firms in these industries than the average for

Table 3.1 *The formation of new manufacturing firms in East Anglia, 1971–81*

Local authority district	Number of new firms	Firm formation rate[1]	New firm employment, 1981
Cambridge	47	4·7	440
East Cambridgeshire	47	21·3	516
Fenland	30	8·0	307
Huntingdon	74	10·2	886
Peterborough	46	2·0	652
South Cambridgeshire	69	8·0	785
Total Cambridgeshire	313	5·7	3,586
Breckland	42	5·2	488
Broadland	14	4·8	93
Great Yarmouth	36	3·5	591
North Norfolk	30	5·1	333
Norwich	42	1·6	407
South Norfolk	17	2·6	181
West Norfolk	27	2·8	455
Total Norfolk	208	3·0	2,548
Babergh	19	3·6	225
Forest Heath	28	6·2	396
Ipswich	19	1·2	366
Mid Suffolk	2	0·5	30
St. Edmundsbury	56	4·6	801
Suffolk Coastal	24	5·3	186
Waveney	34	1·8	340
Total Suffolk	182	2·8	2,344
Total East Anglia	703	3·7	8,478

Note: 1. Firm formation rate: surviving new firms formed July 1971 to June 1981 per 1,000 manufacturing employees in 1971.

new East Anglian enterprises as a whole. Interestingly, these two categories are arguably more characterized by relatively innovative technologies or products than average, such as electronics and computer activities and plastics products and components.

In general, however, the industrial composition of East Anglian new firms plotted in Fig. 3.1 would seem chiefly to reflect the existing industrial structure of the region, together with variations in barriers to entry in particular industries. The latter explains, for example, the

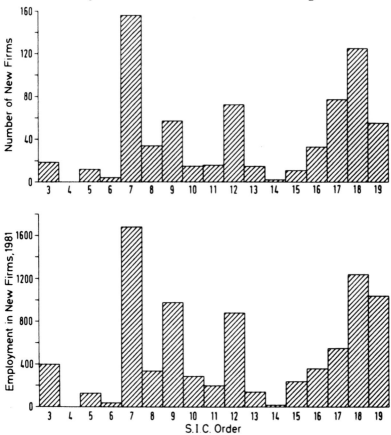

Fig. 3.1 New manufacturing firms in East Anglia, 1971–81: industrial distribution

exceptionally low firm formation rate (Table 3.2) for the region's biggest single manufacturing industry, food, drink and tobacco, where economies of scale in production dictate organization in terms of large canning, freezing or bottling factories. At the other end of the scale, low barriers to entry – and a high turnover of firms – underpin high formation rates in the other metal goods, timber and furniture, and paper, printing and publishing sectors. The high rate (7.3, the third highest of all East Anglian industries) in instrument engineering (order 8) is not however so easily explained in this way.

The comparison of rates of firm formation between different areas of

Table 3.2 *New manufacturing firms in East Anglia,*
1971–81: industrial distribution

SIC Order	Number of new firms	Firm formation rate	New firm employ-ment, 1981	% of total 1981 em-ployment
3. Food, drink and tobacco	19	0·5	403	1·0
5. Chemicals and allied industries	12	1·3	131	1·5
6. Metal manufac-ture	4	3·0	42	2·1
7. Mechanical engineering	156	5·3	1,681	5·8
8. Instrument engineering	34	7·3	344	5·7
9. Electrical engineering	57	2·6	979	6·1
10. Shipbuilding and marine engineering	15	4·2	290	9·7
11. Vehicles	16	0·9	196	1·4
12. Metal goods n.e.s.	72	14·4	841	12·0
13. Textiles	15	4·8	143	7·2
14. Leather, leather goods and fur	2	2·0	16	1·6
15. Clothing and footwear	11	0·9	223	3·2
16. Bricks, pottery glass, cement, etc.	33	4·8	366	5·2
17. Timber, furniture, etc.	77	8·1	546	6·1
18. Paper, printing and publishing	125	7·0	1,242	6·2
19. Other manufac-turing industries	55	5·0	1,045	9·5
Total East Anglia	703	3·7	8,478	4·7

the UK is an essential component of new firms research. This is
hampered, however, by considerable problems with official VAT-
based statistics which in any case do not at present distinguish manufacturing businesses separately (Ganguly, 1982a, 1982b, 1982c), and by
differences between academic studies in data sources, time periods and

Table 3.3 *National comparisons of new manufacturing firm formation*

Area	Time period	Number of surviving new firms	% of end year employment in new firms	Standardized firm formation rate[1]
East Mid-lands[3]	1968–75	1,650	4·2	0·42
Greater Manchester[4]	1966–75	2,312	3·8	0·57[2]
Merseyside[4]	1966–75	553	3·7	0·36[2]
Scotland[5]	1968–77	504	2·2	0·08
Cleveland[6]	1965–78	165	2·8	0·10
Durham[7]	1965–78	236	4·4	0·25
Tyne and Wear[8]	1965–78	486	3·6	0·17
South Hampshire[9]	1971–79	333	2·7	0·34
Cambridge-shire[10]	July 1971–June 1981	313	5·2	0·57
Norfolk[10]	July 1971–June 1981	208	3·5	0·30
Suffolk[10]	July 1971–June 1981	182	3·1	0·28

Notes: 1. Firm formation rate divided by the number of years in the study.

2. Number of new firms substantially overstated: a retrospective telephone check in 1979 revealed that the number of new firms present in 1975 and surviving to 1979 was only 833 in Manchester and 210 in Merseyside.

3. Fothergill and Gudgin, 1979, 1982; Gudgin *et al.*, 1979.

4. Lloyd, 1980; Lloyd and Dicken, 1979.

5. Cross, 1980, 1981.

6. Cleveland County Council, 1982.

7. Durham County Council, 1982.

8. Tyne and Wear County Council, 1982.

9. Mason, 1982.

10. This study.

survey methodologies. It has been possible, nonetheless, to calculate from the latter standardized figures which facilitate direct comparison, with the only caveat relating to the accuracy of some of the studies. The standardized rate is the manufacturing firm formation rate for each area (see above) divided by the number of years covered by the study (Table 3.3). In comparison with other areas of Britain, East Anglia has clearly experienced a high rate of manufacturing firm formation. This is so notwithstanding the possibility of some inflation of estimates for other areas which were not able to be checked by telephone survey. The latter is certainly true for Manchester and Merseyside (see footnote to Table 3.3). It is striking, however, that despite the elimination by the telephone survey of so many apparently new firms in East Anglia, the East Anglian average rate of 0.37 is still high, while the Cambridgeshire rate of 0.57 is markedly greater than in any other part of the country. The comparison with Northern and Scottish rates is particularly stark, while the Cambridgeshire figure is even greater than the East Midlands rate, relating to what has been identified as one of the most successful regions in the UK for new firm formation in the post-war period (Gudgin, 1978).

Although direct comparisons at an international level are even more difficult to make, it is disturbing to note that studies for the USA (Churchill, 1959), Canada (Collins, 1972) and Norway (Wedervang, 1965) indicate that manufacturing entry rates in the UK are probably appreciably lower than in other advanced industrialized countries. This is confirmed by the recent authoritative work on new firm formation in the Republic of Ireland (O'Farrell and Crouchley, 1984) where rates in the 1970s were more than double those of even East Anglia.

TEMPORAL VARIATIONS IN FIRM FORMATION

One of the key questions regarding the extent of new manufacturing firm formation is its ability in the long run to contribute to national economic regeneration. In this connection, some researchers have identified a relationship between national economic performance and levels of firm formation such that periods of recession and increasing unemployment stimulate entrepreneurship by forcing people into setting up their own business following or in anticipation of redundancy (Atkin, Binks and Vale, 1983). Fothergill and Gudgin, 1982, using a sample of 210 surviving companies, have noted this relationship in the East Midlands over the period 1947–75, and show graphically that at least up to 1974, periods of rising national unemployment were closely associated with an increasing number of firm formations. In the context of the current recession work by Binks and Coyne, 1983, argues that this trend is still very much in evidence, although equally high rates of company liquidations may negate the long term impact. While this work does not identify manufacturing firms separately, unpublished research by Gudgin (personal communication) does show a rising trend in total numbers of manufacturing companies in Britain between 1974–81 (from 111,000 to 129,000) as recorded by VAT registrations.

In contrast to these findings the detailed information provided by the 700 East Anglian new firms surveyed directly by the present study does not support the view that in this rural region recession has stimulated a marked surge of firm formation (Fig. 3.2). It is important to note that since the current study deals with surviving new firms, as did Fothergill and Gudgin, 1982, there is an expectation of some inflation of numbers of recently established new firms, since these have had less time in which to close. Notwithstanding this, no recent recession-related increase in firm formation rates is apparent. This is the more interesting in that examination of the pattern before the onset of the post-1973 recession does reveal a relationship between a rising national unemployment rate and an increasing number of new firms. The 1972 peak, then trough, not only coincides exactly with trends in national unemployment, but also parallels precisely the findings of the earlier East Midlands study. However, no relationship is apparent after 1973 when despite a large increase in national unemployment (to 1976), there was no corresponding growth in the number of new East Anglian firms. Indeed, the second highest peak (1977–8) of new firm formation in East Anglia over the ten year period occurred at a time when recession was easing, not intensifying, as indicated by a falling national unemployment rate. Finally, the most recent period of intense recession, 1979–81, reveals no evidence of any further rise in manufacturing firm formation rates in the region. Indeed, the annual rate (surviving new firms in each year per 1,000 manufacturing employees in that year) actually fell, from 0.44 in 1978 to 0.28 in both 1980 and 1981.

The apparent discrepancy between these East Anglian findings and what limited national evidence on rising manufacturing firm formation rates is available is not easy to explain. One obvious possibility is that the East Anglian data is deficient with regard to picking up very recently established companies, who are therefore under-represented in the data bank. Against this, it should be noted that official government 1981 lists were supplemented by current up-to-date commercial directories, Chamber of Commerce lists, and County Planning records, that the telephone survey identified a number of new firms which were *too* recent for inclusion (indicating that the various lists used were up-to-date), and that notwithstanding particularly authoritative data for Cambridgeshire, the post-1979 fall-off in firm formation is as true here

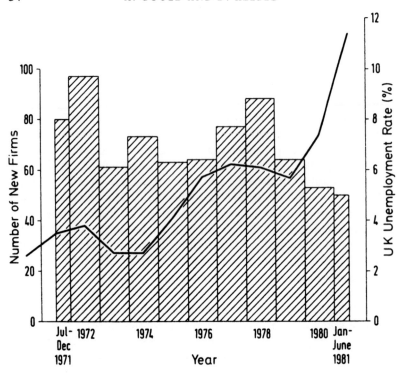

Fig. 3.2 The formation of new manufacturing firms in East Anglia,
1971–81

as in Norfolk or Suffolk. If the data is nonetheless deficient in this
respect, this of course would also imply that the region's already high
firm formation rate is understated in this paper.

   Alternatively, the East Anglian results may be entirely accurate. In
this case, it could of course be that recession-induced firm formation is
not as characteristic of rural regions, with a limited traditional industrial
base and less steeply-rising unemployment, as of the other, far more
urban–industrial, regions of the country, where industrial decline and
unemployment have developed on a much more massive scale. It is also
possible that the rise in the national manufacturing firm population
could reflect in part the establishment of subsidiary companies, rather
than new independent firms. Generally, this position suggests the need
for caution in concluding that recession has stimulated a country-wide
surge of viable new firm births, which may provide a basis for
restructuring British manufacturing industry in the 1980s and beyond.

THE LOCATION AND IMPACT OF FIRM FORMATION WITHIN EAST ANGLIA

There are two dimensions to the spatial distribution of new firm formation in East Anglia. The first, paralleling earlier studies (see earlier section), is a distinct bias towards the rural areas of the region. The East Anglian evidence indicates conclusively that new firms in Britain today are more likely to start outside urban centres and also have greater impact in rural than urban settlements. The urban–rural classification adopted here is based on the population of towns in 1971. Morphologically-defined continuous urban areas have been used rather than administratively designated towns and cities which tend to be underbounded. Quite clearly the major urban centres of Norwich, Ipswich, Cambridge, Peterborough, Great Yarmouth and Lowestoft themselves form a class: they all had 1971 populations in excess of 50,000, together contain about 60% of the region's employment and to varying degrees dominate East Anglia's space economy (Sant and Moseley, 1977). Outside the major urban areas towns with populations greater than 12,000 are indicated by a rank-size distribution to form a distinct group with a clear break in settlement size below this. These medium-sized settlements, in East Anglian terms at least, comprise the small towns category (Table 3.4).

The contrast between firm formation rates in urban and rural areas is striking. The rural areas of East Anglia experienced a firm formation rate of 6.3 which is nearly three times that of the large towns category with 2.2. The small towns rate falls between these extremes. In all 69% of East Anglia's new manufacturing firms have been established in villages or small towns. Moreover, although 1981 Census of Employment data are not yet available, project estimates suggest that the employment impact of new firms is greater in the rural areas, where well over 6% of 1981 manufacturing employment is in new firms compared to only about 3% in the major urban centres and between 4% and 5% in the smaller towns. It seems clear from this contrast that, for whatever reasons, entrepreneurs in East Anglia are selecting locations in the small settlements and rural areas in preference to urban locations.

The second major feature of the spatial pattern of new firm formation in East Anglia is its domination by what may be called 'the Cambridge effect'. In terms of the actual number of firms formed there is a concentration in the south west of the region within about 40 kilometres (25 miles) of Cambridge (Fig. 3.3). This is even more apparent

Table 3.4 *New manufacturing firms in East Anglia,*
1971–81: urban–rural variations

Urban–rural category	Number of new firms	Firm formation rate[1]	New firm employment, 1981
Large Towns[2]	216	2·2	2,592
Small Towns[3]	114	3·4	1,780
Rural Areas	373	6·3	4,106
Total East Anglia	703	3·7	8,478

Notes: 1. See note 1, Table 3.1
2. Cambridge, Peterborough, Norwich, Great Yarmouth, Ipswich, Lowestoft.
3. Huntingdon, March, St. Neots, Wisbech, Kings Lynn, Thetford, Bury St. Edmunds, Felixstowe, Haverhill, Newmarket.

in terms of firm formation rates (Fig. 3.4). All five local authority districts falling in the top quartile of East Anglian firm formation rates – East Cambridgeshire, Fenland, Huntingdon, South Cambridgeshire and Forest Heath – are clustered in this south-western corner, although Cambridge itself has a below average rate. It should be noted that this 'Cambridge effect' is not synonymous with a 'Cambridgeshire effect', since it excludes Peterborough but includes the Forest Heath district of Suffolk.

Explanation of rural bias in East Anglian firm formation is not really possible at the aggregate level of data and analysis presented in this paper. This issue, which is intrinsically important both academically and for local authority planning, is therefore currently under investigation through an interview survey of a large sample of East Anglian new firms stratified by urban and rural origins. However, analysis of spatial variation in formation rates at the level of the region's twenty local authority districts (Fig. 3.4) is feasible, and has been conducted in terms of four possible factors suggested by previous research and knowledge of likely processes of new firm creation. These are the role of existing industrial structure, the size-structure of the existing plant population, the occupational structure of the resident male workforce and the extent of previous industrial in-migration.

The role of an area's existing mix of industries in influencing subsequent industrial change is well recognized (Gudgin, 1978; Cross, 1981). At a local authority district level the combination of industrial structure and the varying propensity of different industrial sectors to

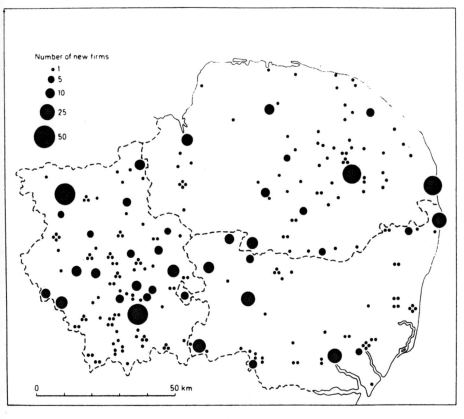

Fig. 3.3 New manufacturing firms in East Anglia, 1971–81

generate new firms may have an important bearing on subsequent district firm formation rates. It is clear from calculating the firm formation rates which would be expected on the basis of each district's industrial structure and aggregate East Anglian firm formation rates for each SIC order (Table 3.2) that this constitutes a partial explanation of variations in actual formation rates across East Anglia. Calculation at SIC order rather than detailed Minimum List Heading level is necessitated by the high frequency of very small or zero samples of new firms and existing industry in many MLHs and districts, and the lack of detailed MLH district employment estimates. On the basis of this calculation, however, the areas of Huntingdon, East Cambridgeshire and Fenland would certainly be expected to be in the upper quartile of

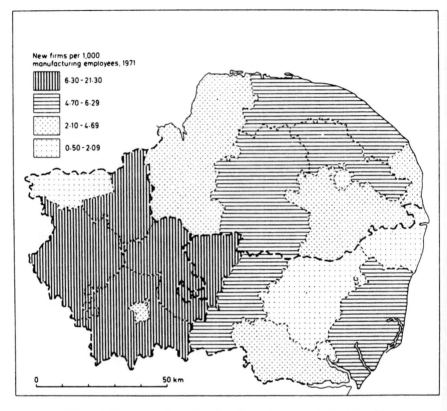

Fig. 3.4 New manufacturing firm formation rates in East Anglia,
1971–81

districts ranked according to firm formation rates: these areas possessed
a mix of industries in 1971 which should have generated a high number
of new firms. But even so, when industrial structure is allowed for, the
'Cambridge effect' is still very much in evidence. As Fig. 3.5 shows,
East Cambridgeshire, Huntingdon, South Cambridgeshire and Forest
Heath still record firm formation rates well above those which would
be expected from their inherited industrial structure. In addition, the
remote but environmentally attractive areas of North Norfolk and
Suffolk Coastal also experienced rates well above those expected.
District variations in 1971 industrial structure thus only partially
explain subsequent variations in firm formation rates.
Perhaps the most widely-accepted influence identified by previous

work on firm formation rates, other than industrial composition, is the size-structure of an area's factories (Lloyd and Mason, 1983, p. 24). Indeed, workers such as Fothergill and Gudgin (1982, p. 132) conclude from careful and extensive analysis of East Midlands and Cleveland data that firm formation rate variations are 'largely' to be accounted for by 'the extent to which local manufacturing employment is concentrated in large plants'. This is because the dependence of an area on employment in such factories tends to suppress local entrepreneurship through not providing the relevant work experience necessary for entrepreneurial training and encouragement. Two strands of evidence support this line of argument. In the first place Gudgin *et al.*, 1979, for the East Midlands; Johnson and Cathcart, 1979, for the Northern region; and Cross, 1981 for Scotland all show that new firm founders are most likely to have worked for a company employing less than ten workers immediately prior to starting their own business. Secondly, Fothergill and Gudgin, 1982, have shown that those towns of the East Midlands with over 50% of their employment in plants of 500 or more people had firm formation rates between 1968–75 only one third those of the other, by definition 'small plant', East Midlands towns.

In some contrast to these arguments, however, the East Anglian evidence does not support the view that local plant size is the key determinant of firm formation rates within the region. A product-moment correlation of district firm formation rates and the percentage of 1971 employment in plants of less than 100 shows only a weak positive relationship (for nineteen districts $r = +0.42$, significant at the 0.1 level). The expected negative relationship between firm formation rates and the percentage of employment in large plants of 500 or more is weaker still (for nineteen districts $r = -0.20$, not significant). In both cases, coefficients improve slightly when plant size is correlated with firm formation rates adjusted for industrial structure (that is, the residual generated by subtracting 'expected' rates based on 1971 industrial structure from 1971–81 rates – see Fig. 3.5). But the gain is not large (e.g. to $r = +0.48$ in the less than 100 plant size analysis, $n = 19$), while the logic of standardizing for industrial structure is perhaps arguable, given the undoubted association between local industrial composition and local plant size distribution. In the wider context of subsequent findings on other factors associated with firm formation rate variations, it must therefore be concluded that in the East Anglian case, plant size structure seems to be only of secondary, not primary significance in influencing local firm formation rates; while to the

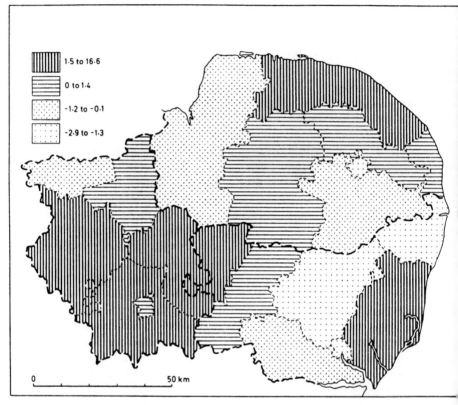

Fig. 3.5 New manufacturing firm formation rates in East Anglia, 1971–81: actual minus expected

extent that it does so, it is the proportion of small plants rather than of large factories which is important.

What other factors, then, determine formation rates in East Anglia? The most obvious alternative to plant size as a controlling influence would seem to be the occupational structure of the resident workforce, especially the male workforce. It has been shown by several researchers that the managerial expertise and educational qualifications of potential founders have a direct bearing both on their propensity to start firms and also on the subsequent success of ventures (Cross, 1981; Storey, 1982). It seems likely that the higher the proportion of the population with managerial, professional, technical and other non-manual qualifications (Lloyd and Mason, 1983, p. 24), the larger will be the pool of potential and potentially successful entrepreneurs. This is viewed by Fothergill and Gudgin, 1982, p. 122 as a secondary influence on firm formation rates, producing local slight variations around norms dictated by plant size. However, in the East Anglian case Fig. 3.6 reveals that there is a much higher positive correlation between firm formation rates adjusted for industrial structure and the percentage of the resident 1971 male population in non-manual occupations than was the case with plant size. Thus for nineteen districts, omitting the abnormally high East Cambridgeshire value produced by an exceptionally small 1971 base year employment, the correlation coefficient is +0.77, significant at the 0.01 level. This is by far the strongest spatial association identified by the East Anglian analyses, and reflects a much stronger initial relationship with unadjusted formation rates ($r = +0.62$, significant at the 0.1 level). Of course, the question of what in turn determines the local socio-economic mix of the resident workforce of an area is an important further issue. In the East Anglian context, factors such as the nature of the local economic base, including the balance between agriculture, manufacturing and services, the possibility of commuting by workers in higher level occupations from the south (and especially south-west) of the region to jobs in London and the northern Home Counties, and the undoubted relative attractiveness of particular areas – in and around Cambridge, North Norfolk or coastal Suffolk (see Fig. 3.5) – may all play a part. Indeed, this could of course be the chief explanation for *rural* bias in new firm formation within the region and elsewhere, if rural areas have tended disproportionately to attract managers and higher income workers for reasons of residential amenity and the perceived benefits of living in historic villages and attractive countryside. But whatever the reasons, the main findings of the formation rate analyses pinpoints local residential occupational structure as the chief – and powerful – determinant of recent rates of new manufacturing firm formation in East Anglia.

The final analysis raises the possibility that a secondary influence on firm formation rates is the impact of previous levels of industrial in-migration to East Anglia. In the period 1966–74 a recorded total of 286 manufacturing establishments moved into East Anglia, primarily from London and the rest of South East England, providing nearly 16,000 jobs by the mid-1970s (Keeble, 1980, ch. 7). Two-thirds of this movement was to the expanded towns of Bury St Edmunds, Haverhill, Huntingdon, Kings Lynn, Mildenhall, St Neots, Sudbury and Thetford. Such moves have been dominated by complete transfers (66% of total East Anglian migrant firm employment, 1966–74) rather than

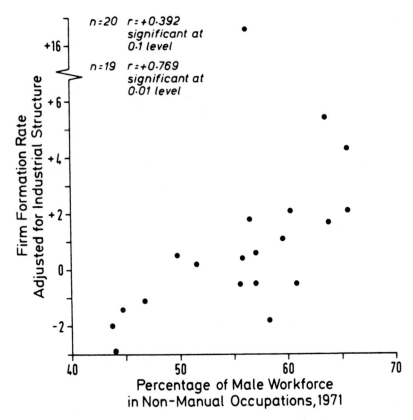

Fig. 3.6 Occupational structure and firm formation rates in East Anglia

branch plant establishment, while most transfers have been of small or medium-sized firms. The possibility therefore arises that such in-migration may have introduced a considerable number of potential firm founders, from growth industries (Keeble, 1976, ch. 6) and with appropriate experience for entrepreneurship, to particular localities: and that over time, this process may in turn have engendered high firm formation rates. One problem with testing this hypothesis, of course, is that much of the impact of earlier industrial in-migration is obviously subsumed within the other 1971 measures – of industrial and plant size structure, and occupational mix – analysed above. Nonetheless, a product-moment correlation between the industrial in-migration rate (the number of migrant firms 1966–74 per 1,000 manufacturing employees in 1971) and firm formation rates (omitting the extremely high value recorded in East Cambridgeshire) produces a coefficient of +0.45 which is significant, as with the plant size measure, at the 0.1 level. Although clearly weak, the relationship noted here is actually stronger than for the factory size variable, suggesting that this may even be a more significant secondary influence on firm formation rates in the East Anglian case.

Overall, then, the aggregate analyses of relationships between different environmental and structural influences suggested by previous work and knowledge of firm generation processes on the one hand, and firm formation rates on the other, provide considerable support for the view that in a rural region such as East Anglia, the chief single determinant after local industrial composition of rates of new firm formation is the occupational structure of the resident workforce. At the same time, secondary influences may well encompass both the influence of earlier industrial in-migration – implying, of course, a continuing feedback process of local and regional industrial growth initiated by such migration – and the role of plant size structures, especially in terms of the local frequency of small factories and firms.

HIGH-TECHNOLOGY NEW FIRM FORMATION

A topic of considerable contemporary interest is the extent to which new firms are being formed in high-technology industrial sectors characterized by high levels of research and innovation, the introduction of new products and processes, and sales and output growth. While some workers have claimed that new firms play only a minor role in technological development (Little, 1977), others such as Rothwell, 1982, p. 366; Rothwell and Zegveld, 1982; and Oakey, 1983, p. 62, argue that many new firms today are technologically innovative and hence important for job generation and national and regional industrial restructuring. Moreover, Hall, 1981, and Thomas, 1983, have claimed that new firms in 'sunrise' high-technology industries in Britain are particularly being generated in locations within 'that broad belt that runs from Oxford and Winchester through the Thames valley and Milton Keynes to Cambridge' (Hall, 1981, p. 537). This vision of a new Krondratieff cycle of high-technology manufacturing growth in the 1980s postulates close access to university research and highly qualified scientists, and high residential amenity for entrepreneurs and researchers, as key influences on high-technology new firm formation and growth. To what extent does recent new firm formation in East Anglia conform to this kind of popular scenario?

Table 3.5 *High-technology new manufacturing firms in East Anglia*

Minimum List Heading	Number of high-technology new firms	1981 employ-ment
271 General chemicals	2	22
272 Pharmaceutical chemicals	2	31
354 Scientific and industrial in-struments and systems	26	253
364 Radio and electronic com-ponents	9	77
365 Broadcast receiving and sound reproducing equip-ment	12	201
366 Electronic computers	13	330
367 Radio, radar and electronic capital goods	8	157
Total East Anglia	72	1,071

The definition of high-technology industry is not easy. Indeed, Beaumont, 1982, p. 2, argues that 'no precise definition of high technology is possible', any definition being time specific and a matter of consensus. Nonetheless, recent work by Kelly, 1983, section 3, has pinpointed nine Minimum List Headings of the Standard Industrial Classification which can be recognized as high-technology on the basis of national industrial characteristics (Table 3.5) notably: R & D expenditure as a percentage of output; rate of technological innovation; and labour force bias towards administrative, technical and R & D workers. In fact, two of Kelly's high-technology MLHs are missing from this table: MLH 363, Telephone and Telegraph Apparatus and Equipment; and MLH 383, Aerospace Equipment, recorded no new firm formations in East Anglia during the study period. While the definition of high-technology firms is MLH-based, the detailed product description provided by each firm in the course of the telephone survey was checked to see if it genuinely accords with the definition in terms of technological sophistication. All firms in these MLHs readily passed this check. The possibility remains, however, that technologically sophisticated firms exist in addition to those recognized. The following statistics could therefore understate the significance and impact of high-technology firm formation in East Anglia.

Of the 703 surviving new firms in East Anglia, seventy-two (10%)

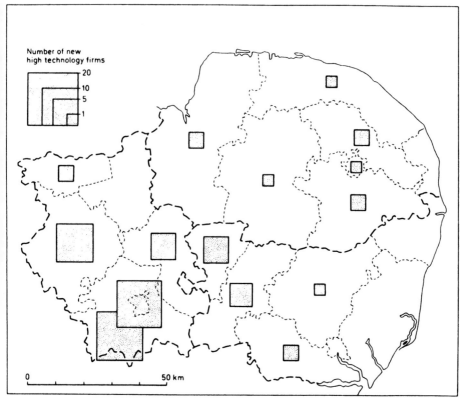

Fig. 3.7 New high-technology manufacturing firms in East Anglia, 1971–81

are high-technology new firms and employ 1,071 people, or 13% of total new firm employment. Thus, high-technology new firms tend to be larger than other new firms with a mean employment size of fifteen and a median of twelve, compared with ten and seven respectively for all other new firms. In terms of their industrial composition, high-techno-logy new firms are clearly dominated by the scientific instruments and electronics industries. Electronics firms alone account for forty-two (58%) of the total, and tend to be even larger than average for all high-technology firms. Average employment in new electronic com-puter firms is no less than twenty-five workers.

Of greatest interest perhaps is the locational concentration of high-technology new firms in and around Cambridge (Fig. 3.7): 75% of these firms have been formed within thirty miles of Cambridge. In other words the 'Cambridge effect' noted for all East Anglian new firms is most strikingly evident in the case of these high-technology firms. Explanation of this effect by statistical analyses of the kind reported previously is not possible in view of the very small size of the total population of firms. There is nonetheless a broad relationship, as with all new firms, with local occupational structure in the form of the proportion of non-manual workers in the resident labour force, and with an inherited industrial structure in the Cambridge region biased towards scientific instruments and electrical engineering. A more complete qualitative explanation is however likely to lie in a complex of actual and perceived environmental advantages for high-technology industry, which are rooted in the labour, entrepreneurial and information advantages of the Cambridge region, the historic presence of Cambridge University as a world-famous centre of scientific research, and the residential attractiveness of this particular part of East Anglia (Gould and White, 1968, p. 172), coupled perhaps with the prestige of a Cambridge post-code. In turn, to the extent that these Cambridge-focused new firms involve new industries, use new technologies, employ highly-skilled workers and generate quite considerable local income multipliers in terms of service jobs and related non-manufacturing activity such as computer software houses (Levi, 1982), this growing cluster of young high-technology enterprises is already of some importance as one, though by no means the only, component in the Cambridge region's relatively buoyant, low unemployment and high income economy. The present study thus does provide some factual support, in terms of the spatial clustering around Cambridge and above-average growth of high-technology new firms, for current hypotheses linking the development of new firms and new technologies to amenity-rich, university-focused areas of southern England.

CONCLUSIONS

The first-stage results of the East Anglian new firms project thus both corroborate and question findings of earlier work in this field. Such work has, of course, been chiefly directed towards urban areas and differences may possibly reflect the focus of the present study on new firm formation in a rural region, characterized by numerous villages and small market towns, a relatively attractive residential environment, and only limited traditional manufacturing industry. Equally, present findings may require modification in the light of the large-scale second-stage interview survey of East Anglian new firms and entrepreneurs currently being analysed by the project team.

Four findings based on the aggregate records of the East Anglian new firms data bank are worthy of special note, particularly with regard to UK government policy, whether central or local, towards new manufacturing firms. First, the East Anglian data do not support the view that recent recession has stimulated a marked surge of new enterprise in the region, at least measured in terms of surviving, and hence reasonably viable, manufacturing firms. While this finding could reflect deficiencies in the data bank, it could also, and equally, bear out the scepticism expressed by Lloyd and Mason, 1983, over the significance of apparently rising national rates of manufacturing firm formation. Certainly the East Anglian finding does not support the view that small firm policies are likely to be a major force in national industrial regeneration in the 1980s, particularly when the very small share of East Anglian regional manufacturing employment in new firms (only 4.7% or 8,500 jobs in 1981) is taken into account. If this is all that can be achieved in a region with one of the highest rates of manufacturing firm formation in Britain, national-level new firm policies seem likely to be of only strictly limited significance, at least in the short term, for industrial restructuring and job generation.

The second finding deserving emphasis is that, as hypothesized by previous workers, such as Fothergill and Gudgin, 1982, p. 121, and Storey, 1982, p. 195, East Anglia does indeed in practice exhibit one of the highest rates of new firm formation so far identified in the United Kingdom, and markedly higher than in assisted regions such as Northern England and Scotland. The rate for Cambridgeshire alone is exceptionally high. While no specific inter-regional analyses to explain this difference are attempted in this paper, the results of the district-level analyses of variations in formation rates within East Anglia are of course consistent with the view that the region's relative abundance of new manufacturing firms reflects four main factors. These are inherited industrial structure, the favourable occupational structure of the resident workforce, small average plant size, and previous complete transfer in-migration by small or medium-sized manufacturing firms. The occupational structure factor may be of particular significance, if the district-level results are any pointer to wider inter-regional influences. From a regional policy perspective, at least, the implications of these findings are somewhat depressing. Not only are national small and new firm policies, to the extent that they are successful, most likely to enhance entrepreneurship and new firm job creation in already relatively better-off regions such as East Anglia, thus widening existing regional differentials in this respect; but also, regional new firm policies aimed specifically at the assisted low formation rate regions seem likely to face major difficulties in narrowing these differentials, if they are indeed rooted in existing regional occupational structures, plant size characteristics, and so on. Even the in-migration finding works against the assisted regions, whose substantial earlier manufacturing migration gains were of course dominantly of branch plants, not transfers as in the East Anglian case (Keeble, 1976, p. 139). Regional new firm policies for the assisted regions, though arguably essential, must surely therefore be seen in a very long term context, with little likelihood of quick or easy results.

The third East Anglian finding of note follows from the above. Within East Anglia there are marked district and urban–rural variations in firm formation rates, favouring the Cambridge region in particular (the 'Cambridge effect') and rural areas in general. While rural bias remains to be investigated in the second stage of the project's work, the latter's district-level analyses spotlight local occupational structure, along with industrial mix, as the key determinants of spatial variations in formation rates within the region, with plant size and industrial in-migration playing a secondary role. The discovery of a markedly stronger sub-regional relationship between firm formation rates and occupational characteristics, than between formation rates and plant size, is of course original to the present study, and to some extent challenges earlier judgements on the relative importance of different determinants. Perhaps the most obvious policy implication involved here relates to planning and rural development policies within East Anglia. Thus the present study indicates that new firms are of greater importance locally, in particular villages or small towns, than aggregate regional statistics suggest, and by inference provides support for local government or Development Commission 'pump-priming' activities, perhaps through the provision of small nursery units or the refurbishing of old farm buildings, aimed at rural enterprise and job creation.

The last conclusion concerns the role of high-technology new firms in East Anglia's industrial development. This role must not be over-inflated. Most of East Anglia (fourteen out of twenty districts – Fig. 3.7) possesses insignificant numbers of such firms, while even in the 'Cambridge region', high-technology new firm employment is only at present a very small component (800 jobs) in the local labour market. Nonetheless, the striking spatial clustering of such firms in the

Cambridge area is noteworthy, while there is also no doubt that these new firms form an integral part of a wider recent high-technology phenomenon here. This is symbolized by, though not dependent on, the development of the Cambridge Science Park (Cane, 1982), and includes the substantial growth of related non-manufacturing research and development activity, computer software firms, externally (and indeed internationally) controlled high-technology subsidiary companies, and formerly independent new firms now taken over by larger organizations. Moreover, high-technology new firms in the Cambridge area are in industries, such as electronic computers, which are growing at the present time, employ highly-paid workers, and generate above-average local income and service job multipliers. Their long-term significance for the economic prosperity of the Cambridge region is thus arguably much greater than their present employment alone might suggest. Current local government policies of site and premises provision for high-technology firms, as with the Cambridgeshire County Council's initiative in developing a 96,000 square-foot high-technology centre adjacent to its Shire Hall, Cambridge, headquarters, would thus seem entirely appropriate as a logical response to the high-technology 'Cambridge effect' documented by this study.

Acknowledgements

The research reported in this article was funded by the Social Science Research Council under contract HR 7402. The authors gratefully acknowledge this support, without which the research would not have been possible. They would also like to thank Jamie Thompson, John Offord, Ray Evans and Tim Kelly for their help at various times during the project.

REFERENCES

Allen G. C. (1961) *The Structure of Industry in Britain.* Longman, London.
Atkin T., Binks M. and Vale P. (1983) New firms and employment creation, *SSRC Newsletter* **43**, 22–3.
Beaumont J. (1982) The location, mobility and finance of new high-technology companies in the UK electronics industry, unpublished research report, Department of Industry, South East Regional Office.
Binks M. and Coyne J. (1983) The birth of enterprise, Hobart Paper 98, Institute of Economic Affairs, London.
Cameron G. C. (1973) Intra-urban location and the new plant, *Pap. Reg. Sci. Ass.* **31**, 125–44.

Cane A. (1982) The Cambridge Science Park: ten year gamble begins to pay off, *Financial Times*, 1 December, p. 14.

Churchill B. C. (1959) Rise in the business population, *Survey of Current Business Magazine*, May.

Cleveland County Council (1982) Manufacturing employment change in Cleveland since 1965, Report 221, Cleveland CC, Middlesbrough.

Collins L. (1972) *Industrial Migration in Ontario*. Statistics Canada, Ottawa.

Cross M. (1980) New firm formation and regional development: the case of Scotland, 1968–1977, unpublished PhD thesis, University of Edinburgh.

Cross M. (1981) *New Firm Formation and Regional Development*. Gower, Farnborough, Hants.

Durham County Council (1982) Manufacturing employment change in County Durham since 1965, Durham CC, Durham.

Firn J. R. and Swales J. K. (1978) The formation of new manufacturing enterprise in the central Clydeside and West Midlands conurbations, 1963–1972, *Reg. Studies* **12**, 199–214.

Fothergill S. and Gudgin G. (1979) The job generation process in Britain, Research Series 32, Centre for Environmental Studies, London.

Fothergill S. and Gudgin G. (1982) *Unequal Growth: Urban and Regional Employment Change in the UK*. Heinemann, London.

Ganguly A. (1982a) Births and deaths of firms in the UK in 1980, *British Business*, 29 January, pp. 204–7.

Ganguly A. (1982b) Regional distribution of births and deaths of firms in the UK in 1980, *British Business*, 2 April, pp. 648–50.

Ganguly A. (1982c) Regional distribution of births and deaths of firms in the UK, *British Business*, 24 September, pp. 108–9.

Gould P. R. and White R. R. (1968) The mental maps of British school leavers, *Reg. Studies* **2**, 161–82.

Gudgin G. (1978) *Industrial Location Processes and Regional Employment Growth*, Saxon House, Farnborough.

Gudgin G., Brunskill I. and Fothergill S. (1979) New manufacturing firms in regional employment growth, Research Series 39, Centre for Environmental Studies, London.

Hall P. (1981) The geography of the fifth Krondratieff cycle, *New Society*, 26 March, pp. 535–7.

Howick C. and Key T. (1979) *Small Firms and the Inner City: Tower Hamlets Gatsby Project on Local Authority Economic Planning*. Centre for Environmental Studies, London.

Johnson P. S. and Cathcart D. G. (1979) New manufacturing firms and regional development: some evidence from the Northern region, *Reg. Studies* **13**, 269–80.

Keeble D. E. (1976) *Industrial Location and Planning in the United Kingdom*. Methuen, London.

Keeble D. E. (1980) East Anglia and the East Midlands, in Manners G., Keeble D., Rodgers B. and Warren K. *Regional Development in Britain*, pp. 177–200, Wiley, Chichester.

Kelly T. (1983) The location, organisation and regional impact of high-technology industry in the United Kingdom: proposals for a PhD research programme, Department of Geography, Cambridge University.

Levi P. (1982) Cambridge: the place where success breeds growth, *Financial Times*, 30 November.

Little A. D. Ltd (1977) *New Technology-Based Firms in the UK and the Federal Republic of Germany*. Wilton House Publications, London.

Lloyd P. E. (1980) New manufacturing firms in Greater Manchester and Merseyside, Working Paper 10, North West Industry Research Unit, University of Manchester.

Lloyd P. E. and Dicken P. (1979) New firms, small firms and job generation: the experience of Manchester and Merseyside, 1966–1975, Working Paper 9, North West Industry Research Unit, University of Manchester.

Lloyd P. E. and Dicken P. (1982) Industrial change: local manufacturing firms in Manchester and Merseyside, Inner Cities Research Programme 6, Department of the Environment, London.

Lloyd P. and Mason C. (1983) New firm formation in the UK, *SSRC Newsletter* **49**, 23–4.

London Industry and Employment Group (1979) *The Potential for Local Enterprise: A Study of the Furniture Industry in London*. LIEG, Middlesex Polytechnic.

Macey R. D. (1982) Job generation in British manufacturing industry: employment change by size of establishment and by region, Government Economic Service Working Paper 55, Regional Research Series No. 4, Department of Industry, London.

Mason C. M. (1982) New manufacturing firms in South Hampshire: survey results, Discussion Paper 13, Department of Geography, University of Southampton.

Mason C. M. (1983) Some definitional difficulties in new firms research, *Area* **15**, 53–60.

Nicholson B. and Brinkley I. (1979) The inner city incubator hypothesis and the development of new manufacturing firms, Department of the Environment project (mimeo).

Oakey R. P. (1983) New technology, government policy and regional manufacturing employment, *Area* **15**, 61–5.

O'Farrell P. N. and Crouchley R. (1984) An industrial and spatial analysis of new manufacturing firms in Ireland, *Reg. Studies* **18**, 221–36.

Rothwell R. (1982) The role of technology in industrial change: implications for regional policy, *Reg. Studies* **16**, 361–9.

Rothwell R. and Zegveld W. (1982) *Innovation and the Small and Medium Sized Firm: Their Role in Employment and Economic Change*. Frances Pinter, London.

Sant M. and Moseley M. J. (1977) *The Industrial Development of East Anglia*. Geo Abstracts, Norwich.

Storey D. J. (1982) *Entrepreneurship and the New Firm*. Croom Helm, London.

Storey D. (1983) How beautiful is 'small'? *SSRC Newsletter* **49**, 18–20.

Thomas D. (1983) England's Golden West, *New Society* **64**, 5 May, pp. 177–8.

Tyne and Wear County Council (1982) Manufacturing employment change since 1965, Tyne and Wear CC, Newcastle upon Tyne.

Wedervang F. (1965) *Development of a Population of Industrial Firms*. Scandinavian University Books, Oslo.

Wood P. A. (1974) Urban manufacturing: a view from the fringe, in Johnson J. H. (Ed.) *Suburban Growth – Geographical Perspectives at the Edge of the Western City*, pp. 129–54. Wiley, London.

4

Spatial variations in new firm formation in the United Kingdom: comparative evidence from Merseyside, Greater Manchester and South Hampshire

P. E. LLOYD and C. M. MASON

INTRODUCTION

From the very limited evidence available, it appears that the number of new firms started each year in the UK has been rising since 1974, and has continued to rise during the present recession (Bannock, 1981; Binks and Coyne, 1983). This fact has been cited as evidence of success by a Conservative Government seeking to promote the creation of a new 'climate of enterprise' in order to 'encourage people who would like to have a go and build up their own business' (John MacGregor, Under Secretary of State for Industry, *Daily Telegraph*, 5.10.81). In reality, however, such statistics present a picture of trends in the new firm sector which at best is deficient and which may even be seriously misleading. The increasing number of new firms has, for example, been accompanied by a sharp rise in numbers of company deaths (Bannock, 1981; Woods, 1982; *British Business*, 1983). While total numbers of company registrations have been slowly increasing in recent years, the *rate* of registration has been virtually constant (Woods, 1982).[1] A large proportion of the new firms started in recent years have been in the volatile service sector. Government statistics indicate, for example, that in the period 1980–2 inclusive new *manufacturing* firms accounted for only 10% of all new businesses started during these years (Ganguly, 1983). While not regarding companies created in one sector as in any way inferior to those created in any other sector, it is nevertheless likely that the proportion of new businesses that are simply replacements rather than genuine additions is larger in the service sector than in manufacturing.

One aspect of the deficiency of aggregate statistics on new business formation which is of particular concern is that they tend to disguise the nature of the *spatial distribution* of new business enterprise. The UK space-economy is by no means equally well endowed either to generate or to benefit from new firm formation over all of its sub-regions. A picture which displays the aggregate conditions for new firm initiation and development as being derived from a relatively uniform situation from place to place and which serves to deflect attention from the variety of conditions which prevails in reality can indeed be seriously misleading.

NEW MANUFACTURING FIRM FORMATION IN THE UK: THE SPATIAL DIMENSION

It is a feature of the social background to new firm formation that very few founders move their homes in order to set up a new business. Instead, most new businesses are located close to their founders' places of residence (Gudgin, 1978; Scott, 1980; but see Cross, 1981, for a dissenting view). Any subsequent re-location conditioned by the growth of new enterprises is similarly highly localized (Cross, 1981). Consequently, therefore, spatial variations in new firm formation and the subsequent growth of such enterprises will tend to exist as a product of the different propensity of particular areas to generate new firms and see them grow. As Cross, 1981, had suggested, one of the fundamental variables conditioning such spatial variations in new firm initiation and development is the nature of the local labour market as a supply-side factor determining the potential availability of new firm founders.

In general terms, new firm formation tends to be depressed in those regions which specialize in traditional heavy industries, especially where a small number of large plants dominate the local labour market (Chinitz, 1961). The products which are often based upon 'one-off' designs clearly themselves offer little scope for small firm production, while the scale of start-up capital requirements imposes a significant barrier to entry. Emphasis in such industries is on production rather than marketing, the management style is frequently paternalistic, corporate organization is often old-fashioned and consequently there is a slow adoption of modern management methods with little demand for specialized management skills and limited opportunity for occupational mobility by managers. Production technologies emphasize traditional craft skills and demand strength and endurance from factory-floor employees rather than problem solving abilities. Here too there is often only limited opportunity for individual initiative and advancement (Segal, 1979).

In such areas as these, therefore, local populations have neither the opportunity and incentive nor do they develop the skills needed to set up new businesses. Such a line of argument has been effectively used by Checkland, 1981, to explain why, during the twentieth century, West Central Scotland failed to generate indigenous new firms able to offer some diversification for the regional economy. He suggests that the traditional concentration on shipbuilding and heavy engineering created a milieu in which other kinds of activities were unable to take root. Using the analogy of the legendary upas tree which was believed to have the power to destroy other growths for a radius of 15 miles under its shade, Checkland argues that 'the upas tree of heavy engineering killed anything that sought to grow beneath its branches' (p. 12). The same line of reasoning might equally well be applied to other heavy industrial regions in the UK such as North East England and South Wales. Despite the continued contraction of these heavy industrial complexes, their influence on local entrepreneurial climates remains powerful. Moreover, it is possible to argue that upas trees of more recent growth may be represented by those branches and assembly-line factories whose dissemination under the auspices of regional policy has taken them to precisely those areas where a previous tradition of heavy industry has already stunted indigenous enterprise. While hard evidence on this issue is difficult to come by, some recent work does show a clear relationship between an area's plant size structure and its rate of new firm formation. Employees who work in small firms are, it seems, more likely to set up a new business than those working in large firms (Johnson and Cathcart, 1979a; Fothergill and Gudgin, 1982). Since many of the areas which have a large plant size structure are in the peripheral regions of Britain and most areas with a higher share of small plants are in the Midlands and South (Fothergill and Gudgin, 1982; Lever, 1979), it seems not unreasonable to anticipate that inter-regional rates of new firm formation in the UK will decrease with movement north and westwards.

While the size of an enterprise may itself have an impact on the propensity of its workforce to acquire the skills and attitudes essential for small business initiation and management, this represents only one of a whole cluster of correlated variables which influence the nature of work experience and its appropriateness for the varied demands levied by small firm management. Skilled manual workers are, it is suggested, better equipped than unskilled and semi-skilled workers for small firm entrepreneurship because they acquire more of the problem-solving skills required, while management employees, particularly where they have had some responsibility for financial matters or some involvement with marketing and sales, seem likely to be better equipped than manual workers to start a business, though not necessarily to turn out as good a product. In the broader context, therefore, another of the general variables conditioning the supply potential of suitable small firm entrepreneurs is the occupational spectrum of a city or region. Where the majority of the enterprises are small, the occupational spectrum (the balance of management to skilled and production-line workers) will tend to reflect this (Storey, 1982), reinforcing the observed propensity for traditional small firm areas to have good rates of new firm formation. Where, however, the local industrial base is dominated by large corporate enterprises, the mix of skills may well be antithetic to the requirements for entrepreneurship. The spatial division of labour within multi-site enterprises has, for example, resulted in many peripheral regions being dominated by externally-owned branch plants performing routine assembly operations with a truncated range of management functions. Within the UK, multi-site companies tend to concentrate their headquarters and divisional offices and, perhaps more important, their R & D laboratories in the South East and the adjacent regions of East Anglia and the South West. Even single-site companies in these regions employ a larger proportion of such staff than their counterparts in the rest of the country (Crum and Gudgin, 1977). As a result, the proportion of the regional population in those managerial, professional and technical occupations most suited to small business entrepreneurship is highest in the South East (Fothergill and Gudgin, 1982; Storey, 1982). This South East bias is further reinforced by the tendency for multi-site enterprises to manufacture new products at their headquarters location and generally close to their R & D laboratory, whereas products reaching the end of their life cycle tend to be the responsibility of peripheral plants (Oakey *et al.*, 1980). The stable technology associated with mature products is, in many ways, unlikely to provide a stimulus for the secondary spin-off of new technology-based firms (Thwaites, 1978). In contrast, the concentration of R & D laboratories in the South East (Buswell and Lewis, 1970) would tend to suggest that numbers of high technology spin-offs will be substantially higher in this region than elsewhere in the country.

Among other conditioning variables, the growth of new enterprises appears to be strongly related to the background and education of the founders (Storey, 1982). Firms started by those with a management background, particularly if they have a degree or professional qualification, show the fastest rates of growth (Fothergill and Gudgin,

1982). Often prior work experience has rendered them aware of sources of outside finance and of the conventions necessary in presenting successful cases for loan funds. At a personal level, many will have accumulated adequate collateral against which loans can be secured (Cross, 1981). By contrast, many new businesses which are started by those with basic education and manual rather than professional backgrounds display low rates of growth, not least because of the limited aspirations of their founders, their lack of personal capital and their frequent reluctance to use outside sources of finance. Together these tend to produce chronic undercapitalization and a dependence on subcontract work from a limited range of customers. In many respects, therefore, the base map of socio-economic and occupational status for the regions of the UK will, itself, offer a surrogate both for the supply of entrepreneurs and for the degree of expected success in the average case.

Such differences in creditworthiness and of access to personal and institutional finance will feed forward to condition levels of launch and post-start-up capital as a product of the personal attributes of founders. Higher returns from both second mortgages and from the use of the domestic home as collateral for a bank loan has the effect of raising the threshold of personal capital availability in those regions with relatively higher levels of owner occupancy and with higher housing values. For regions like Northern Ireland, Scotland, Northern England and Yorkshire-Humberside the low level of owner occupancy (Central Statistical Office, 1982) denies many of their putative small businessmen the collateral base from which many might be able to induce banks to advance risk funds. Currently, however, redundancy lump sum payments may be performing this role for those who choose to invest them in their own enterprise rather than in secure savings funds.

Most new firms, from a demand-side perspective, serve local and regional market areas (Johnson and Cathcart, 1979b; Storey, 1982). Relatively few first-time enterprises are set up on the basis of a product of their own and most are engaged in sub-contract work for larger companies and institutions (Gudgin, 1978). On both counts, therefore, the rate of new firm formation and the subsequent growth of such enterprises will tend to be significantly influenced by the level of final and intermediate demand in the local and regional economy which itself will rest upon the performance of corporate 'prime-movers' and public sector agencies. It has been suggested (Lloyd and Dicken, 1982) that this relationship can be conceptualized as a 'crowded platform' – a platform of finite dimensions which is served by two staircases. One provides the route for new firms to join the platform, the other provides room for the newcomers to be accommodated as displaced firms leave. Broadly, it can be proposed that the 'market room' (on the platform) for newcomers is provided by their capture of markets vacated either voluntarily (by retirement) or involuntarily (out-competed) by those local firms further ahead of them in the development sequence. In broader terms, it might be argued that local and regional 'platforms' will differ in scale over the cycle and hence in their ability to accommodate new entrants. For example, because of higher general income levels in South East England and as a result of the *relative* prosperity of its key electronics and defence industries, there may be greater market opportunities for the new firm than in peripheral industrial areas such as North East England and West Central Scotland. These latter areas, with their low income levels and contracting shipbuilding and heavy engineering industries, have tended to experience a decline in those sub-contracting opportunities available for new and small firms (Rabey, 1977; Gibb and Quince, 1980). Similarly, the growing scale of branch-plant activities in their economies may further exacerbate this situation as outsourcing produces little locally articulated demand to stimulate the growth of local small enterprise (Lever, 1974; McDermott, 1976; Hoare, 1978; Marshall, 1979).

Much of the evidence ranks as circumstantial but there would seem to be a case *a priori* that on account of its socio-economic and industrial structure South East England (especially those parts of the region which lie outside Greater London), together with parts of the adjacent regions of East Anglia and the South West, will exhibit the highest rates of new firm formation and may be expected to have the highest proportion of rapidly growing and innovative enterprises. By contrast, northern industrial regions, and particularly the conurbations, will display low rates of new firm formation and survival. Storey's attempt to devise an index of regional entrepreneurship adds further weight to this conclusion (Storey, 1982). But those attempting to reduce the degree of speculativeness on questions of inter-regional variations in new firm formation face a serious lack of hard and reliable evidence. Certainly, government data on new firm formation do provide some confirmation of the hypothesized pattern. For example, the only regions with rates of new firm formation above the national average in 1980 were East Anglia and the South East (Ganguly, 1982). However, these data cover a single year only. Moreover, all sectors, apart from agriculture, are included and it is not inconceivable that some variation in new firm formation, especially in the service sector, is directly related to changes in regional population size. The Department of Trade and Industry Record of Openings and Closures confines its coverage to the manufacturing sector but has two most serious weaknesses: first, the omission of firms with less than eleven employees (the employment cut-off is twenty workers in Greater London and fifty in the West Midlands conurbation); and, second, the inclusion of both newly formed enterprises and new manufacturing units established by previously non-manufacturing enterprises. Interpretation of these statistics is, therefore, extremely problematical. Evidence on spatial variations in innovative new enterprises is even more scanty. However, adding locational information (where possible) to Little's list of new technology-based firms started since 1950 (Little, 1976) confirms that such enterprises are largely found in southern regions: over half were located in the South East (in most cases outside Greater London). In all, the South East, South West and East Anglia together accounted for almost three-quarters of such firms.

THE STUDY AREAS: BACKGROUND CONDITIONS FOR NEW FIRM FORMATION

While the range of case-study areas to be addressed was by no means chosen with the requirements of this paper in mind, Merseyside, Greater Manchester and South Hampshire do provide interesting contrasts in those background conditions which the previous section of the chapter suggests should produce some variations in the quality and quantity of new firm start-ups. In the case of Merseyside, for example, there is the archetypal branch plant economy in a depressed peripheral region. Over 70% of Merseyside's (manual) manufacturing workforce was employed in large plants with more than 500 workers in 1975 and only 14% was under the control of locally headquartered enterprises (Lloyd and Dicken, 1982).

Structurally, there is little evidence, within the manufacturing sector, of those industries with a known propensity to spin-off small enterprise. Food processing, chemicals and electrical and mechanical engineering on Merseyside tend to favour those particular parts of the manufacturing process which, these days, generate only limited local multipliers (Lloyd and Cawdery, 1982). Even a long-standing tradition of small firm sub-contracting to the Birkenhead shipyard failed (for reasons which Rabey, 1977, showed in the case of the North East) either to survive the demise of the prime customer or to diversify into new markets. In addition, the vehicles industry, so widely canvassed in the early sixties as a potential source for local small firm sub-contract opportunity, failed to generate the local multiplier effects foreseen by its government sponsors. Outside manufacturing in the narrowly defined sense, the construction sector offers some market pull to the local entrepreneur but in its typically cyclical fashion. Merseyside remains dominated by service sector employment as a residue of its heritage as a port. Indeed, only 36% of the conurbation workforce was classified as being in manufacturing in 1977. Here too the opportunities for non-manufacturing small firms in the transport and distribution field, though once considerable, have now been considerably reduced on the demise of the Port of Liverpool.

In occupational terms Merseyside's industrial history has left a legacy of bias toward the manual element of the skill spectrum for the bulk of the active population, with a heavy representation of the semi-skilled and unskilled whose work experience is unlikely to have provided an impetus toward entrepreneurship. Further, as an area with a depressed living and working environment suffering from an exaggerated image of poor labour relations, Merseyside would seem unlikely either to attract an inward movement of potential entrepreneurs or to retain those local entrepreneurs whose activities give them freedom of locational choice in siting their production facilities.

Therefore, *a priori*, there is an expectation that rates of new firm start-up in manufacturing on Merseyside would be low despite the considerable activities of central and local government in sponsoring new enterprise in the area. Over the period 1971–7 the local economy suffered continuing job loss as a result of retrenchment among the larger employers. Its total employment declined by 11.5% but the workforce in manufacturing fell by in excess of 20%. The impact of recession after 1979 has yet to be adequately quantified but rising redundancy and accelerated closure rates (again by the larger enterprises) has drawn employment down to a level where unemployment stands at over 18% (August 1983). Thus, while the demand for some form of new enterprise on Merseyside has never been greater, the logic of the background conditions for new firm formation would seem in this case to suggest that the supply of such firms is likely to be seriously limited.

Greater Manchester, though only 30 miles from Merseyside and co-located within what is currently one of England's most depressed regions, presents a sharp contrast. Here those background conditions which are assumed to have an important bearing on new firm formation and survival rates are more positive. As the first industrial city,

Greater Manchester retains a tradition for indigenous enterprise that long pre-dates the growth of the multiplant, multi-locational firm. Nineteenth century institutions which provided support to Manchester-based firms, though considerably depleted from their heyday, continue to provide an infrastructure of support for local enterprise. The small and medium sized firm still retains a substantial stake in the local economy with only 37% of manufacturing companies employing more than 500 workers in 1975. Though in general decline, indigenous (locally-headquartered) enterprise still accounted for almost half (48%) of all manual employment in the conurbation at that date (Lloyd and Dicken, 1982).

In occupational terms Greater Manchester offers a history of traditional manufacturing skill among both men and women primarily associated with its heritage as a textiles and engineering area. In environmental terms also the *Cottonopolis* heritage remains strong, with the empty or partially occupied mill or inner city warehouse continuing to provide a prominent element in the urban fabric and a source of cheap, if run down, premises for the small firm start-up.

The apparent advantages offered by Greater Manchester for new firm formation and small firm survival must, however, be discounted against the disadvantages which flow from the sectoral distribution of the bulk of the local small firm population. For the most part the traditions of local entrepreneurship have remained closely tied to the textiles-clothing complex and to those engineering activities which service it. This complex has, of course, been in decline for over half a century but perhaps never at such a rate as at present with the effects of a newly emerging international division of labour and a world recession striking at its roots.

In the case of Greater Manchester, therefore, in sharp contrast to that of Merseyside there does exist a tradition of small firm entrepreneurship and a supporting business and financial infrastructure. Such local small enterprise is, however, closely associated with an industrial complex in rapid decline and is set within an urban fabric which bears witness to the scale of that decline over several decades. While entrepreneurship skills may be assumed to exist within the working population and while there is widespread indigenous experience of working in the small firm sector, these attributes remain closely attached to the textiles-clothing complex and are not necessarily readily transferable to other forms of local enterprise.

Although its industrial demise is less widely publicised than that of Merseyside, Greater Manchester continues to suffer serious losses in its manufacturing employment base. Between 1971–7, the conurbation suffered a loss of almost 16% in its industrial job stock. Only the compensatory growth of service employment, now at an end in recession, served to offset the aggregate impact exerted by closure and redundancy in the manufacturing sector (Lloyd and Reeve, 1982). Unemployment in Greater Manchester currently stands at over 14% (August 1983) and here, as on Merseyside, great emphasis is being placed upon the small, indigenous firm as a source of replacement job generation – a prospect perhaps more favourable owing to its longer tradition of small firm entrepreneurship but which has to be tempered by the sectoral bias in its present stock of indigenous enterprise.

In terms of broad contrasts in the background conditions for small firm establishment and survival, the case of South Hampshire offers substantially different attributes from those of the two North West conurbations. This area, which comprises the Southampton and Portsmouth city-regions and the New Forest, has been one of the fastest growing sub-regions in the country since 1945. Total employment increased by 10% between 1971–7 compared with 2% nationally, and while much of this growth was provided by service industries, the manufacturing sector increased its workforce by 3% at a time of general decline for the nation as a whole. The onset of recession since 1979 halted employment growth in South Hampshire but, even then, its level of redundancies has been running at about half the national rate. The rate of unemployment is, therefore, below the national average, at 11.4% in the Portsmouth travel-to-work area (TTWA) and 8.3% in the Southampton TTWA (August 1983).

South Hampshire does not possess a strong entrepreneurial heritage and prior to the Second World War its industrial base was extremely narrow. The Portsmouth economy was dominated by the Royal Naval Dockyards while that of Southampton was derived from its function as Britain's premier ocean-going passenger port. Moreover, the small numbers of major industrial employers in each city were primarily non-local enterprises (Riley and Smith, 1981; Ford, 1934; Mason, 1981). The post-war industrial diversification and expansion of South Hampshire occurred largely as a result of the in-movement of considerable numbers of companies, for the most part decentralizing from London. Not only did their arrival swamp the contribution made by indigenous enterprise to local economic development but it also produced a situation in which the area is currently dominated by large, externally-owned establishments. For example, plants with 500 or more employees accounted for 47% of total manufacturing jobs in 1979 and around three-quarters of total manufacturing employment was provided by non-local firms. Nevertheless, the area does possess an indigenous small firm sector which is significant in numerical if not in employment terms. Moreover, externally-owned establishments in South Hampshire seem to bear little resemblance to those in peripheral regions. The latter are characterized by their tendency to offer a truncated range of employment opportunities and functions. By contrast, many of the externally-owned establishments in South Hampshire are represented either by a divisional head office or by the headquarters of the UK subsidiary. A considerable number of the larger non-locally-owned establishments also undertake on-site R & D activities and are frequently responsible for products at an early stage in their life cycle. These features are reflected in an occupational structure in which the full range of managerial, technical and skilled manual workers is well-represented. South Hampshire remains, however, an area with only limited dependence on industry with manufacturing accounting for around 31% of total employment in 1978.

In the terms set out in the introduction to the paper, South Hampshire's industrial structure offers further advantages for new firm formation and growth. Approximately one in every five manufacturing jobs are, for example, in the electronics industry and a further 5% of jobs are provided in other parts of the electrical engineering sector. Given that new firm founders generally start their businesses in the same industry in which they previously worked, South Hampshire would seem particularly well placed for the formation of new high technology enterprises. In addition, the major electronics companies in the area, along with firms in its sizeable mechanical engineering and shipbuilding industries, continue to present a major market for small firms undertaking sub-contract work.

The entrepreneurial base of South Hampshire has been further enhanced by the considerable in-migration of population to the area during the past two decades or so, a function of the combined attractions of plentiful job opportunities and the widely held perception that it offers an attractive residential environment. To a large degree, population migration is selective, with young families, skilled manual workers and management staff having the highest propensities to migrate. A large proportion of entrepreneurs are drawn from the ranks of such groups in the population. Thus, South Hampshire – along with other parts of the outer South East – has received a large in-migration of that sub-set of the economically active population which might be expected to have a high propensity for the formation of new enterprises (Keeble, 1978).

Despite, or possibly because of, its lack of industrial tradition, South Hampshire offers, therefore, an environment for new firm formation which differs substantially from that of the two North West conurbations. The in-migration of potential new firm founders, relatively buoyant market opportunities for companies undertaking sub-contract work, the expansion of the local economy and the presence of an industrial workforce which is mainly engaged in problem-solving rather than routine tasks are all likely to promote a high level of local entrepreneurship.

In general terms, therefore, the three areas whose new enterprises are explored in what follows offer sharply constrasting conditions for the initiation and development of local enterprise. Judgements, *a priori*, might tend to suggest that in relative terms, the 'sunbelt' conditions of South Hampshire would give it a greater propensity both for new firm formation in general and specifically for the creation of new rapid-growth enterprises. Greater Manchester, with its long-established small firms tradition, while having some anticipated advantages for new firm formation, might have this tempered under contemporary conditions by the sectoral bias of its economy and the currently depressed fortunes of the textile and clothing industries. A similar 'trade-off' might also exist for Merseyside with background conditions providing low expectations for new enterprise formation but with an active policy environment favouring experimental new firm start-up.

Aggregate data upon which to test such broad assumptions are, as we suggested in the introduction, virtually impossible to come by. With the assistance of the Department of Trade and Industry, however, we have been able to construct Table 4.1, which examines the distribution of enterprises new to manufacturing (ENMs) in each of the three areas in the periods[2] 1966–75 and 1976–80. Stocks of enterprises against which to measure the newcomers were not available for the base years but, to give some scaling factor for the three areas, the recorded numbers of manufacturing establishments for the appropriate TTWAs for 1972 and 1978 (mid-years) were utilized.

In the event, the record for 1966–75 seems, clearly, to indicate the expected high propensity of South Hampshire for new enterprise start-up with the two northern conurbations lagging far behind in relative terms and showing little variation between themselves. The period of weak recovery and recession after 1976, however, reveals uniformly low inter-regional levels of new firm formation with South

Table 4.1 Enterprises new to manufacturing (ENMs) for the survey areas 1966–75 and 1976–80

	ENMs 1966–75	Manufacturing plant stock 1972	Formation rate	ENMs 1976–80	Manufacturing plant stock 1978	Formation rate
Merseyside[1]	43	1,212	3·5	38	1,161	3·3
Greater Manchester[1]	87	3,935	2·2	51	3,738	1·4
South Hampshire[2]	70	532	13·1	7	596	1·2

Source: Department of Trade and Industry: special tabulation
Notes: 1. TTWAs approximating Metropolitan County.
2. Portsmouth and Southampton TTWAs only.

Hampshire having added only a handful of newcomers to its substantial earlier influx.

Of course, the Department of Trade and Industry data on ENMs do have serious shortcomings. In particular, they exclude firms with less than eleven employees; given that most new firms employ very few workers, at least in their early years, the effect of this cut-off is that the majority of new enterprises in any area are omitted from the data (as an example, see Nicholson *et al.*, 1981). A further problem, which operates in the opposite direction, is that included within the ENM statistics and accounting for about 15% of the total are new manufacturing establishments set up by previously non-manufacturing firms (Pounce, 1981). The data have also been criticized as deficient in identifying new firms that satisfy the criteria for inclusion (Johnson and Cathcart, 1979b), a point which the Department of Trade and Industry has seemingly acknowledged as valid for the 1976–80 period (Killick, 1983).

Given these deficiencies, it would be unwise at this stage to read too much into the aggregate data. They are presented here only as a broad indication of relative trends and have the merit that they do at least illustrate the overall more favourable position of South Hampshire. Indeed, because the data are confined to the Portsmouth and Southampton TTWAs and exclude the New Forest area which had the highest rate of new manufacturing firm formation in South Hampshire during the 1970s (Mason, 1982), this favourable position may even be understated. The size and relative stability of the Merseyside new firm share *vis-à-vis* Greater Manchester, which is also indicated by the data, might provide some weak support for the operation of a policy effect. (The importance of policy in the perception of Merseyside's small firms generally was clearly illustrated in Lloyd and Dicken, 1982.) Softer data for the three areas can be used to buttress the impressions derived from aggregates and it is to these that we now turn.

QUALITATIVE CONTRASTS AMONG NEW MANUFACTURING FIRMS IN THE THREE STUDY AREAS

The new firm surveys

The basis for the more qualitative analysis of new manufacturing firms which follows rests upon survey work conducted by the authors over the period 1980–1. In the case of Greater Manchester and Merseyside, this was carried out under the auspices of research contracts from the

Department of the Environment and the Department of Industry, the results of which are published in full in Lloyd and Dicken, 1982. In the case of South Hampshire, the work was carried out under SSRC sponsorship and results are available in Mason, 1982.

New firms research has its own special methodological difficulties not least of which is that of defining the new enterprise itself, since the main elements in any definition – start-up date, independence and newness – are themselves open to different interpretations (see Mason, 1983a). Moreover, since, as Brinkley and Nicholson, 1979, have shown, many young business enterprises undergo a transition from service activity to manufactured output, any definition based upon their activities as observed at a single point in time may not be sufficiently flexible. With these issues in mind, new firms had to satisfy three main criteria to be eligible for inclusion in each of the survey programmes. First, only firms which had begun trading during or after 1976 were included. Second, the surveys were restricted to those firms which, though they may have started in another form, were engaged in manufacturing at the survey date. Finally, only founder-based enterprises were eligible for inclusion. In these, the founder–owner or owners accounted for the bulk of the company equity at the time of survey.

The surveys also had to grapple with the problem of deriving comprehensive and comparable lists of all new enterprises in each area. The best available source of information on new firms was the identical establishment data-banks (based on the Health and Safety Executive records) covering North West England and South Hampshire and which, at least in theory, included all manufacturing establishments regardless of size. Extensive telephone screening was undertaken to ensure that each new enterprise identified from this source satisfied the start-up date, independence and manufacturing criteria. Randomly selected quota samples of thirty new firms were drawn in each of Merseyside and Greater Manchester (as part of a wider survey covering some 460 indigenous enterprises in all) but in the case of Merseyside this quota exhausted the entire population of firms available for interview. In South Hampshire, a total of sixty-two enterprises were identified as satisfying the survey criteria and the owner-managers of fifty-two of them agreed to be interviewed (an 84% response rate). Essentially the same questionnaires were used in each area.

Given this basis, it is acknowledged that the overall findings in the subsequent sections must still be regarded as somewhat speculative in nature. Nevertheless, it is believed that by offering some comparative case-evidence which uses identical data sources, agreed definitions and a common questionnaire the paper does make some useful contribution to the debate. In what follows, there is an attempt to summarize the findings rather than to present a fully worked out comparison. Indeed, rigorous analysis is precluded by the inevitable problems in this kind of research in acquiring relatively 'hard' comparable information about the operation of new enterprises. We begin by examining the characteristics of the founders of the surveyed firms in the three areas and then move on to consider some of the features of the enterprises themselves; in each case, the aim has been to establish the nature of the similarities which were in the ascendency and to point up the key contrasts.

The founders of new enterprises in the study areas

In general terms, new firm founder-owners in the manufacturing sphere were remarkably similar between Greater Manchester, Merseyside and South Hampshire. For the most part, consistent with the picture presented by the literature, around two-thirds of founder-owners had backgrounds within manufacturing and, in particular, within the small firm sector. In terms of previous employment around one-third of the new entrepreneurs had acquired their work experience in firms with more than 500 workers. Again consistent with emerging evidence from other studies, the bulk (around 80%) of the founders in all three areas were local residents or had worked locally immediately prior to setting up their firms. Also consistent across the studies was the widely observed tendency for the initiation of the business to rest on the combined efforts of two or more co-founders. Often one would remain in full-time employment while the other launched the enterprise. Start-up was itself by no means a discrete event. Instead, many newcomers 'edged' their way toward independent, full-time, self-employment over a number of months or even years. As to motivation, it is perhaps tautological to cite a 'desire for independence' as the key but, in this respect, there was great emphasis in all three areas for the desire to be one's own boss to be the prime motivation for start-up.

The contrasts between the regional cases, where they did emerge, were individually small but cumulatively important. First, founder-owners in South Hampshire tended to be *significantly older* than their Merseyside and Manchester counterparts. While 57% of the new entrepreneurs in each of the two North West conurbations were in the 30–45 age range (a similar proportion to that reported for Scotland by Cross, 1981), only 53% of the sampled founders in South Hampshire fell into this age group. The difference lay in the proportion of founders over 45 in South Hampshire (32%) as compared with Merseyside (19.1%) and Greater Manchester (10.6%).

Consistent with this finding is the other key contrast represented by the proportion of new firm owners with previous experience in founding an independent enterprise. Whereas only 10.6% of the founders on Merseyside and 25.5% in Greater Manchester had *previous experience* of founding an independent business, the equivalent proportion in the South Hampshire sample was 42%. Given the critical importance of market awareness to the success of small business (Lloyd and Dicken, 1982), it is also significant that in responding to questions about motivation some 60% of new founders in South Hampshire referred to their desire to exploit a perceived *market opportunity* while the equivalent proportion on Merseyside was as low as 10% and in Greater Manchester only 19%.

In summary terms, then, we begin to see some important differences between founder-owners in the three sampled areas. Slightly older (and perhaps more creditworthy), slightly more experienced in business formation as a group and slightly more conscious of the marketing issue, the founder-owners of South Hampshire (almost half of whom had moved into the area for job-related reasons and had not been born there) seemed to have a marginal propensity as a group to be more successful, though the average individual was probably little different from his northern counterpart. Within the two contrasting North West settings, there is evidence of further important differences. On Merseyside, there were relatively few young new business starters in the set but even those of more mature years had virtually no previous experience and little consciousness of the marketing aspect of new business ventures.

The financial context

As with the characteristics of new firm founders, it seems that the average new enterprise in all three areas presents a broadly similar reflection of the financial attributes typical of the new enterprise. Contrasts appear at the margin with, as we shall argue, the South Hampshire sample of new firms different only in respect of the proportion of better performers at the top of the range.

At the launching stage of the business, which is often (as we suggested earlier) a long-drawn out affair, there is considerable variation in the scale of capital investment. While the median launch capital sum quoted (with all the attendant difficulties of quantification) was £5,000–£6,000 across the three areas, many enterprises were started with less than £500. For the most part, the latter were skilled tradesmen undertaking sub-contract work on their own account and requiring little more than basic power tools, with the customer providing the materials on free issue. Generalization over what was a highly skewed distribution of launch capital proved virtually impossible, though once again there was some hint that the South Hampshire cases were consistently marginally better endowed with launch capital than were their northern counterparts.

With regard to the sources of launch capital, the same phenomenon appears. While the mix of sources is relatively similar – a heavy dependence on personal resources (savings, support from family and friends) with secondary recourse to bank loans and overdrafts – the proportion of responses naming the bank as a source of launch capital is sharply different. In each of the two North West conurbations, only 15% of new starters reported having sought loans or overdrafts. The equivalent proportion for the South Hampshire respondents was 42%.[3] Once again *awareness, experience, and creditworthiness* could be cited as a pre-condition consistent with what we know of the sample of founder-owners.

In examining other financial indicators for the new firms, there were inevitable difficulties of deriving consistent and relatively accurate variables. In the event, the level of financial information available even to the new firm owners themselves was extremely small and, in general terms, limited to a single consistent value for gross turnover. Observed variations in reported profit or returns on capital invested were far more a function of poor measurement, the variability of definitions and general (often unfounded) optimism than of real differences in small firm performance.

Taking gross turnover as the only reliable indicator, the median case on Merseyside reported a value of £67,000, that in Greater Manchester £113,000 and in South Hampshire £120,000. It, therefore, seems clear that, whereas there is little general difference on this variable to separate Manchester and South Hampshire, the Merseyside new starters were significantly smaller. Closer inspection of the individual cases does, however, reveal some contrast between Manchester and South Hampshire with the upper quartile of firms in the latter area *turning over more business* than their equivalents in the former, even though the median and lower quartile values are broadly similar.

On other indicators, the pattern of overall similarity but of regional contrasts at the upper margin of performance is maintained. In general terms, for example, less than 5% of the new firms in all three areas owned the freehold of their premises and almost half of their plant and equipment had been bought second hand. Post start-up investment was generally modest with only £8,000 to £10,000 having, on average, been spent on plant and equipment and around £1,000 on premises by the new enterprises. Only at the upper quartiles of the distribution were substantial differences observable. In respect of plant and equipment stock valuations, for example, only 12% of the Manchester and 8% of the Merseyside firms had plant stocks valued at more than £75,000 compared with 31% for South Hampshire. Similarly, the upper quartile value for post start-up investment in the southern sample was £36,000 for plant and equipment and £4,000 for buildings compared with £20,000 and £2,000 respectively for the combined sample from the northern conurbations.

The tentative conclusion to emerge from what is endemically patchy and far from adequate data on the varying financial context confronting the groups of new firms is that, while the average cases are perhaps little different in their attributes, the South Hampshire sample reveals the presence of a rather more heavily capitalized, higher turnover group of firms at the upper tail of the distribution. It is, perhaps, the absence of such a leavening of the more dynamic enterprises which renders the prospects for small firms on Merseyside and in Greater Manchester so depressing by comparison.

A second feature to emerge from this tentative evidence, by no means necessarily unconnected with the first, is the apparently greater willingness of the South Hampshire founder-owners to gear themselves to commercial lending sources. Causality here is, however, hard to assign. Could it be simply that being more successful in terms of their product-markets they are more successful in gaining access to favourable loan terms or could the causation run in the reverse direction? Access to loan and venture capital finance might itself be a function of the relative creditworthiness of the entrepreneurs and could, therefore, provide the resources and 'elbow-room' necessary to give new starters a better opportunity both to explore market needs and to develop good products.

Products and customers: the market context

With respect to industry type, the new firms in the older conurbations continue generally to belong to traditional 'metropolitan' industries in the jobbing engineering, metal goods, clothing, printing and furniture trades. In the case of South Hampshire, in the absence of a strong metropolitan industry tradition, there is a more significant bias toward engineering and metal processing (42% of the surveyed firms). Significantly, however, around 12% of the South Hampshire firms were classified as electronics-based whereas the North West conurbations had only one firm in this sector. The average firm in all three study areas appears consistently to be the 'one-off' producer performing jobbing work or small batch manufacturing, filling a limited market gap usually as sub-contractor to a small number of key customers.

Despite the presence of firms described as producing electronics products or services, it would be wrong to characterize any of the groups as containing a substantial proportion of technologically advanced enterprises. In the two North West conurbations, only ten firms in total claimed to be engaged in 'innovative' forms of production. The equivalent number for South Hampshire was twelve. The firms' own claims were in general, however, a function of some rosy-hued views about their own capacity. The true proportion of innovative concerns has to be considered as representing little more than a handful in each case. From this evidence it would appear there is little to suggest that the propensity of the South Hampshire firms to represent a more sophisticated group at the upper tail of the performance distribution is derived from the possession of a more up-market structure in terms of product-market or technological competence. However, there is a constrast between surveyed firms in terms of their competitive strengths. Whereas price was ranked by firms in Manchester and Merseyside as by far their most important competitive advantage, firms in South Hampshire gave much greater emphasis to their non-price strengths, notably the flexibility and service provided and the quality of their product or service.

Characteristically, new firms in all the study areas were primarily engaged in serving local markets. Almost three-quarters of the new firms on Merseyside conducted over half of their business within the local conurbation, while the equivalent proportions in Greater Manchester and South Hampshire were more similar at 41% and 50% respectively. There were, however, some exporters in each area with 28% of responses in Greater Manchester and 25% in South Hampshire claiming some form of overseas sales. Consistent with a greater degree of local market dependence, only 7% of the Merseyside firms sold products overseas. On balance, however, the market opportunity available to new firms appears to be strongly based upon what the local community has to offer in the form of demand and here, too, there is a possible factor conditioning the marginal propensity of the southern group to perform better. Available industrial and consumer demand within the relatively prosperous south, even in these hard times, would seem to offer a slight widening of the market-base underpinning the 'crowded platform' available to accommodate local enterprise. While allowing more local firms to survive, greater local final demand may also offer the opportunity for each to have a slightly higher threshold of operation by comparison with a similar set of firms in a depressed northern conurbation.

Environmental considerations: the role of premises

It has become an established feature of policy initiatives for the promotion of small, new firms to favour supply side initiatives primarily by such environmental means as the creation of small workshop factory units, sheltered environments and industrial improvement areas. All seem to be predicated upon an assumption that the ˙environmental-premises constraint has traditionally been a serious device for holding back the formation of new business enterprise in the United Kingdom. Indeed, a government sponsored study (Coopers, Lybrand-Drivers Jonas, 1980) has looked into the problem with a view to guiding developing policy. It has to be said, however, that in the case of the Greater Manchester–Merseyside study the environmental constraint appeared to be considerably less severe than those founded upon difficulties of market penetration, the problems of gaining access to venture capital and the genuine shortage of entrepreneurship. The responses of established new (and young) firm founder-owners in the two northern conurbations indicated that from their perspective the provision of premises and the state of the physical environment had little weight as a *pre-condition for start-up*. While it may be argued that such a survey cannot measure the latent demand for premises by those frustrated in their ambitions to start new enterprise, the current surplus of industrial floorspace at all sizes would seem to qualify such criticism. Only 7% of the firms surveyed experienced real practical difficulties in finding adequate premises, though many more anticipated some difficulty *should they wish to expand their productive capacity*. Short leases at low rents in premises at around the 4,000 square foot mark were those attributes most in demand from new firms and there seemed to the respondents in the two cities no real shortage of premises to fit these needs.

In the case of South Hampshire, the broad nature of demand seemed relatively similar with firms looking for rented premises close to home with short leases to minimize the length of their, still uncertain, commitment. However, with a median floorspace of 3,100 square feet (and quartile values of 1,500 square feet and 5,000 square feet) there was some sign that the size of premises demanded (or at least ultimately obtained) by new enterprises in South Hampshire was below that of their counterparts in the two North West conurbations where the median value was 4,000 square feet (with quartile values of 2,000 square feet and 8,000 square feet).

In marked contrast to the North West, however, and perhaps as a reflection upon the geographical origins of those perspectives which stress the premises constraint, finding suitable premises *did* prove to be a problem for the majority (70%) of new firms in South Hampshire. The reasons for this contrast are threefold. First, in comparison with the North West, South Hampshire was largely by-passed by industrial development in the nineteenth century and so it does not have a large stock of old industrial premises available at low rents and on short leases to act as a 'seedbed'. Second, the scenic parts of the sub-region have, at least until recently, been subject to strict anti-industry planning regulations and this has both restricted the development of new industrial sites to a small number of locations and has also prevented new firms from occupying redundant farm buildings or other empty non-industrial premises. Third, unlike Merseyside, and to a lesser extent Manchester, local authorities in South Hampshire have until very recently shown little interest in building small premises, and because it is not an assisted area, there has been no Department of Industry involvement in the construction of small units. Differences in the industrial backgrounds of the study areas are reflected by the fact that whereas 52% of firms in South Hampshire occupied pre-1940 premises and just under one quarter occupied nineteenth century property, the equivalent proportions in the North West were considerably higher at two-thirds and 38% respectively. However, in this respect there are substantial contrasts between the two North West conurbations. Industrial development policy by both central government and the local authorities has provided Merseyside with a stock of small, new subsidized premises and these have attracted substantial numbers of new firms. Greater Manchester, in contrast, had made little headway in providing nursery units by the time of the survey and its new firms were generally housed in the oldest and most marginal low-cost, short-lease stock made available by the private sector.

Once the problems of finding suitable premises were overcome, most new firms in South Hampshire seemed generally satisfied that their immediate requirements were being met. As in the North West, rent, security of tenure, physical conditions and the suitability of the premises for current activities were regarded by the majority of firms as satisfactory whereas lack of space for expansion and the suitability of the premises for future activities were perceived as poor.

On balance, therefore, there seems far more substance to the contrast between the North West conurbations and South Hampshire in respect of environmental provision than in many other issues explored by the survey. Such a contrast may, in broader terms, be seen to justify what, from a viewpoint from the older urban areas of Britain, seemed an unaccountably heavy emphasis by policy-makers on the premises-sites variable. With a growing propensity (Fothergill and Gudgin, 1982) for new enterprise in general, apparently, to seek out ex-urban sites in previously non-industrial environments, the South Hampshire case is likely to represent just one example of the operation of a premises constraint and the unwillingness of the planning system readily to accede to the new trend in rural areas. It is in such instances that an *aspatial* view of the new firms issue can be at its most misleading. The supply-side constraints of the northern conurbations singly or in unison are by no means inevitably the same as those afflicting the areas of expansion in rural counties, and inter-regional analyses and locally-centred planning are needed to provide spatial/local sensitivity to centrally derived conventional wisdom about the nature of the small firm problem.

Employment and the new firm

Another commonly held view about the new firm in general is that it provides, at a time of large firm restructuring, the most readily available source of labour intensive enterprise with which to offset rising unemployment. To a degree, this is true yet, as the surveys show, the number of new firms of average size required to offset rising redundancy seems far beyond the capacity of the economy to generate in the short term.

In terms of employment, as with so many other attributes, the surveys of Merseyside, Greater Manchester and South Hampshire produce clear consistency. The median employment of surveyed firms in all three areas was, for example, eight employees with the mean standing at eleven for each of the northern conurbations and twelve for

South Hampshire. Against the Bolton criterion of 200 workers for the definition of the small firm, these new enterprises of less than five years' standing are minute and to convert their average size into the numbers required against current employment losses in either area would demand the initiation of *thousands* of them over a short time. Such numbers are clearly not forthcoming.

It might, of course, be argued that out of the normal population of new starters described here the seeds of a new ICI or GEC may emerge to make nonsense of calculations based upon an average size of around ten workers. However, the more extended survey of a representative sample of 460 firms in Manchester and Merseyside (including young and long-established indigenous firms) indicated that survival rates were low for new starters, with 60% failing to survive for ten years. There was also little evidence of the emergence of more than one or two firms able to grow to medium size (Lloyd and Dicken, 1982). This may, of course, represent a feature of a particular population of firms in two depressed northern cities. The 'Silicon Valley phenomenon' may be anticipated as being more likely in South Hampshire but, although the research here is at an earlier stage, there is little evidence to date that new firms represent such a dynamic breed as to produce large numbers of workers in the foreseeable future. Indeed, a follow-up study of the panel of fifty-two firms found that only two had achieved a rapid growth in their employment in the two years since they were originally surveyed (mid-1981 to mid-1983). Moreover, a surprisingly high proportion – almost one-quarter – had closed over this period (Mason, 1983b).

In occupational terms, the firms in each area are again broadly similar with a tendency for the small firms as a group to have a relatively high dependence upon skilled workers (36% in South Hampshire, 37% in Greater Manchester and 33% on Merseyside). There is also a leavening of technical and professional workers in what is often a more balanced structure than the average branch plant or assembly operation. Perhaps significantly, the proportion of technical/professional workers in South Hampshire is three times that for the northern conurbations, standing at 9.5%.

Once again, therefore, in the case of employment, South Hampshire evidences a slight edge over its northern counterparts – a slightly higher technical content perhaps reflecting a difference in the nature of production in the southern firms. Skill shortages continue, however, to be reported as a constraint in each area. While around one half of the firms in the Hampshire survey experienced shortages of skilled workers, some one third of the northern firms claimed to have unfilled vacancies for skilled workers. Such constraints have, however, to be examined closely before a snap judgement is made about the need to train more skilled operatives. For the most part, in the northern city surveys, the clue to the shortage was in the wages issue with small enterprises simply unable or unwilling to offer those wages necessary to attract the workers they wanted.

CONCLUSION

In general terms, perhaps the most remarkable feature of the comparative analysis of new firms as between two depressed peripheral conurbations and an emerging area of the outer South East lies not in the degree of contrast observed but in the *degree of similarity*. On balance, the population of new firms may be expected to be relatively similar as between such areas regardless of differences of industrial structure. The nature of the market niches available to new small firms seems to ensure that initially they follow relatively similar pathways into the business community. The *median* new firm realistically, then, can be visualized as having around ten workers, operating in premises at about the 3,000 to 4,000 sq. ft. mark on short leases with low rents and performing those 'one-off' or small batch type of sub-contract operations that place high demands on individual worker skills. Most appear to turn over around £100,000 gross per year and seek to grow primarily from a base within the limits of retained profits. Most serve the local market and are closely tied to a small number of key customers. Growth prospects are uncertain and death rates within a ten-year span may see more than half of the stock of firms falling into liquidation. Only a mere handful of such firms would normally expect to prosper and though, for these, early growth may be extremely rapid, long-run development sees a slowing down and a shift from market penetration to a more conservative approach to retaining market share. By any stretch of the imagination, therefore, such firms are not made of the stuff of the more extravagant claims of their lobbyists. Case examples of success may be not too difficult to find but set within a context of the small firm population as a whole, the chances of success must be rated as being considerably below those of failure.

With respect to inter-regional contrasts, what the studies seem to show is a marginal favourability on most indicators in the direction of South Hampshire. The contrasts are not individually sharp but collectively they point to a population of new firms which contains slightly older founders, marginally more experience and credit-worthiness and a small bias toward the use of more business skills in seeking the support of financial institutions and in having an awareness of marketing issues. There was little firm evidence that the entre-preneurs of South Hampshire were, as a group, engaged in more definitively up-market activities but there did appear to be a small leavening of those more sophisticated enterprise-makers and boffin businessmen whose firms were likely to grow more quickly.

At the other extreme, Merseyside's new manufacturing enterprises seemed to reveal those attributes which the discussion of broad factors in small firm formation would have led us to anticipate. The firms were smaller in terms of turnover (despite having higher initial valuations of plant and equipment – a possible policy effect), they were operated by founder-owners with less previous experience of business, although this was not simply a function of their youthfulness. They were predominantly tied to demand generated in the local (depressed) market. The wider study (Lloyd and Dicken, 1982) showed them to be well endowed with premises but less sophisticated in business terms.

In the broadest sense, then, the gradation from Merseyside, through Manchester to South Hampshire seems to offer some evidence (albeit on limited numbers of firms and with limited information) for the expected order of differential in both quantitative and qualitative terms. While economic variables in terms of the availability, size and range of market niches available to small businesses clearly have a powerful impact on development, entrepreneurship and the general socio-cultural setting for the new enterprise clearly also play a part.

Acknowledgements
The authors acknowledge the financial support of the SSRC (grant HR 6796) and the research assistance of Colin Taylor in undertaking the South Hampshire survey, and the financial support of the Department of the Environment Inner Cities Directorate and the Department of Industry plus the research assistance of Joan Cawdery, Paul Moran and Andrew Thompson in the North West conurbations study.

NOTES

1 This data source is restricted, by definition, to firms adopting limited liability status. Businesses operating as sole traders or partnerships, thought to be the fastest growing type of business in numerical terms, are therefore excluded.

98   P. E. LLOYD AND C. M. MASON

2 Both ENM and Plant Stock Data record only establishments with more than ten employees.
3 South Hampshire firms also made greater use of external sources of finance for post start-up investment than their counterparts in Manchester and Merseyside.

REFERENCES

Bannock G. (1981) The clearing banks and small firms, *Lloyds Bank Review* **142**, 15–25.
Binks M. and Coyne J. (1983) The Birth of Enterprise, Hobart Paper 98, Institute of Economic Affairs, London.
Brinkley I. and Nicholson B. (1979) New manufacturing enterprises: entry mechanisms, definitions and the monitoring problem, Working Note 542, Centre for Environmental Studies, London.
British Business (1983) Insolvencies in England and Wales, 29 April, pp. 182–3.
Buswell R. J. and Lewis E. W. (1970) The geographical distribution of industrial research activity in the United Kingdom, *Reg. Studies* **4**, 297–306.
Central Statistical Office (1982) *Regional Trends*. HMSO, London.
Checkland S. G. (1981) *The Upas Tree: Glasgow 1875–1975 . . . and after*, 2nd edition. University of Glasgow Press, Glasgow.
Chinitz B. (1961) Contrasts in agglomeration: New York and Pittsburgh, *Am. Econ. Rev.* **51**, 279–89.
Coopers and Lybrand Associates Ltd/Drivers Jonas (1980) *Provision of Small Industrial Premises*. Small Firms Division, Department of Industry, London.
Cross M. (1981) *New Firm Formation and Regional Development*. Gower, Farnborough, Hants.
Crum R. E. and Gudgin G. (1977) *Non-production activities in UK manufacturing industry*, Regional Policy Series 3, Commission of the European Communities, Brussels.
Ford P. (1934) *Work and Wealth in a Modern Port*. Allen and Unwin, London.
Fothergill S. and Gudgin G. (1982) *Unequal Growth: Urban and Regional Employment Change in the UK*. Heinemann, London.
Ganguly P. (1982) Regional distribution of births and deaths in the UK, *British Business*, 24 September, pp. 108–9.
Ganguly P. (1983) Births and deaths of firms in the UK, 1980 to 1982, *British Business*, 8 April, pp. 14–15.
Gibb A. and Quince T. (1980) Effects on small firms of industrial change in a Development Area, in Gibb A. and Webb T. (eds.), *Policy Issues in Small Business Research*, pp. 177–89. Saxon House, Farnborough, Hants.
Gudgin G. (1978) *Industrial Location Processes and Regional Employment Growth*. Saxon House, Farnborough, Hants.
Hoare A. G. (1978) Industrial linkages and the dual economy: the case of Northern Ireland, *Reg. Studies* **12**, 167–80.
Johnson P. S. and Cathcart D. G. (1979a) The founders of new manufacturing firms: a note on the size of their 'incubator' plants, *J. Ind. Econ.* **28**, 219–24.
Johnson P. S. and Cathcart D. G. (1979b) New manufacturing firms and

regional development: some evidence from the Northern Region, *Reg. Studies* **13**, 269–80.

Keeble D. (1978) Industrial decline in the inner city and conurbation, *Trans. Inst. Brit. Geogr.* **3**, 101–14.

Killick T. (1983) Manufacturing plant openings 1976–1980: analysis of transfers and branches, *British Business*, 17 June, pp. 466–8.

Lever W. F. (1974) Manufacturing linkages and the search for suppliers and markets, in Hamilton F. E. I. (Ed.) *Spatial Perspectives on Industrial Organization and Decision-Making*, pp. 309–33. Wiley, London.

Lever W. F. (1979) Industry and labour markets in Great Britain, in Hamilton F. E. I. and Linge G. J. R. (eds.) *Spatial Analysis, Industry and the Industrial Environment*, vol. 1, *Industrial Systems*, pp. 89–114. Wiley, Chichester.

Little A. D. Ltd (1976) *New Technology-Based Firms in the United Kingdom and the Federal Republic of Germany*. Anglo-German Foundation, London.

Lloyd P. E. and Cawdery J. (1982) The Large Manufacturing Firm in Liverpool: Corporate Sector Influences on the Development of Indigenous Enterprise. North West Industry Research Unit, University of Manchester.

Lloyd P. E. and Dicken P. (1982) *Industrial change: local manufacturing firms in Manchester and Merseyside*, Inner Cities Research Programme 6, Department of the Environment, London.

Lloyd P. E. and Reeve D. E. (1982) North West England 1971–1977: a study in industrial decline and economic restructuring, *Reg. Studies* **16**, 345–59.

McDermott P. J. (1976) Ownership, organization and regional dependence in the Scottish electronics industry, *Reg. Studies* **10**, 319–35.

Marshall J. N. (1979) Ownership, organization and industrial linkage: a case study in the Northern Region of England, *Reg. Studies* **13**, 531–57.

Mason C. M. (1981) Recent trends in manufacturing employment, in Mason C. M. and Witherick M. E. (Eds.) *Dimensions of Change in a Growth Area: Southampton Since 1960*, pp. 52–74. Gower, Aldershot, Hants.

Mason C. M. (1982) New manufacturing firms in South Hampshire: survey results, Discussion Paper 13, Department of Geography, University of Southampton.

Mason C. M. (1983a) Some definitional difficulties in new firms research, *Area* **15**, 53–60.

Mason C. M. (1983b) Small businesses in the recession: a follow-up study of new manufacturing firms in South Hampshire, Discussion Paper 25, Department of Geography, University of Southampton.

Nicholson B. M., Brinkley I. and Evans A. W. (1981) The role of the inner city in the development of manufacturing industry, *Urban Studies* **18**, 57–71.

Oakey R. P., Thwaites A. T. and Nash P. A. (1980) The regional distribution of innovative manufacturing establishments in Britain, *Reg. Studies* **14**, 235–53.

Pounce R. J. (1981) *Industrial Movement in the United Kingdom, 1966–1975*. HMSO, London.

Rabey G. F. (1977) Contraction poles: an exploratory study of traditional industry decline within a regional industrial complex, Discussion Paper 3, Centre for Urban and Regional Development Studies, University of Newcastle upon Tyne.

Riley R. C. and Smith J.-L. (1981) Industrialization in naval ports: the Portsmouth case, in Hoyle B. S. and Pinder D. A. (Eds.) *Cityport Industrialization and Regional Development: Spatial Analysis and Planning Strategies*, pp. 133–50. Pergamon, Oxford.

Scott M. (1980) Independence and the flight from large scale: some sociological factors in the founding process, in Gibb A. and Webb T. (eds.) *Policy Issues in Small Business Research*, pp. 15–33. Saxon House, Farnborough.

Segal N. S. (1979) The limits and means of 'self-reliant' regional economic growth, in Maclennan D. and Parr J. B. (Eds.) *Regional Policy: Past Experience and Future Directions*, pp. 212–24. Martin Robertson, Oxford.

Storey D. J. (1982) *Entrepreneurship and the New Firm*. Croom Helm, London.

Thwaites A. T. (1978) Technological change, mobile plants and regional development, *Reg. Studies* **12**, 445–61.

Woods L. (1982) Trends in company 'births' and 'deaths', Henley Management College, Henley (mimeo).

5

An industrial and spatial analysis of new firm formation in Ireland

P. N. O'FARRELL and R. CROUCHLEY

INTRODUCTION

During the 1970s the Irish manufacturing sector has been coming increasingly under overseas control and ownership, primarily through the process of branch plant formation, to reach a level of 34.3% of employment by 1981. Although overseas companies have bestowed a wide range of benefits upon the Irish economy (and it is not argued that efforts to attract more should be re-directed), it nevertheless remains a key strand of Irish industrial strategy to stimulate and expand the indigenous industrial base. Between 1973–81 a total of 2,047 new industrial plants were opened throughout Ireland and *survived* until the end of the period (for definitions, see Appendix 1). These openings comprise 407 multinational branches (MNEs) providing 32,365 jobs by 1981; 158 new subsidiaries of Irish multiplant firms (IMPs) with 7,197 jobs; and 1,482 new indigenous single plant firms (ISPs) employing 18,032. This paper focuses upon the indigenous new firm formation process and presents evidence on temporal trends in formation rates, spatial variations at regional and county level and inter-industry differences; an attempt is also made to analyse new firm formation rates both sectorally and spatially within a multivariate framework in order to identify some of the factors underlying variations in entry. Finally, policy implications of the results are discussed.

DATA

The analysis is based upon the Industrial Development Authority's (IDA) annual employment survey conducted on January 1st each year. The survey constitutes a population census of manufacturing establishments with a minimum payroll of three including owner man-ager(s).[1] In subsequent years, if the employment total falls below three, the plant is retained on file and the employment recorded. Prior to 1979, the employment survey only included a sample of plants in County Dublin employing less than fifty people; but data for all new firms which survived in Dublin is available for 1981. Hence, Dublin must be excluded in any national analysis of *all* new firm entries, including those which subsequently closed, but the capital may be incorporated when the objective is to study survivors only. The following data are recorded for every plant: name of firm; total male and female employ-ees; location; product code; year production commenced; nationality; and programme under which grant aided, if applicable. The ownership variable, so fundamental to any investigation of new openings, is not recorded by the IDA employment survey and, therefore, the relevant information was collected. Ownership has been classified, in the case of surviving plants, according to their status in 1981 and, for plants which closed, their status in the year prior to closure (see Appendix 1). This means that the methodology cannot adequately account for the effects of ownership changes between the base and terminal years (t_1 and t_2). In Britain, changes in organizational status – for example, the takeover of a fast growing single plant indigenous firm (ISP) by a larger multiplant firm – is frequently associated with closure (Dicken and Lloyd, 1978, p. 183). Takeover and merger activity is a highly infrequent phenom-enon in Irish manufacturing and the classification of plants according to their organizational status in 1981 is likely to understate the contri-bution of ISPs to employment change to only an extremely marginal degree.

The identification of new manufacturing firms according to what-ever definition is employed has always been a difficult problem for research workers (Mason, 1983). The definition employed in this Irish study is founded upon the concept of the firm as one which has no obvious parent in any existing business enterprise. This distinguishes between subsidiaries established by existing companies – both domestic and overseas – and new independent indigenous firms. Independence has been defined in legal terms recognizing, however, that many independent firms may be functionally dependent. The staff of the IDA classified the ownership variable and a test of the reliability of their classifications was conducted in 1983 when over 300 indigenous single plant firms were randomly sampled and surveyed. Only 1.5% of firms reported that they were subsidiaries of existing businesses thereby confirming the high degree of accuracy achieved by the IDA in identifying indigenous single plant firms. The date of start-up of the new independent firms is defined as the year of entry to the IDA employment survey (Appendix 1).

Name changes are monitored and recorded when they occur; but a change of nationality or product group is only picked up *immediately* in cases where these changes also involved a name change. Where a change of nationality or product group occurs but the plant's name remains unchanged, this is only recorded if noticed by the IDA regional office and subsequently reported. There is evidence to suggest that such changes within plants between 1973–80 are rare phenomena (Appendix 1).

TEMPORAL TRENDS IN NEW FIRM FORMATION BY TOWN SIZE LOCATION

The number of new indigenous firms being established each year outside Dublin averaged 134 per annum between 1973–5; then, from 1976, the rate of entry rose rapidly to peak in 1978 and fall back slightly with the onset of the recession (Table 5.1). The rate of new firm formation was some 46% higher between 1976–80 than over the 1973–6 period, and this trend is manifest outside the Dublin County area. The second row of Table 5.1 records the number of new firms which survived until 1981 including those in Dublin. It is a less meaningful index of the annual fluctuations in entry rates because plants established in earlier years will have been vulnerable to closure for longer periods. However, it does give an approximate indication of the national trend especially since 1977 when the Small Industry (SI) programme of the Industrial Development Authority (IDA) was extended to the capital city.

Indigenous new firm formation rates are classified by town size group in Table 5.2 which has to be restricted to survivors only in order to include Dublin. Hence, caution must be exercised when interpreting entry rates for earlier years whose cohorts have been in existence for longer. It has been shown elsewhere that closure rates do not vary by town size location (O'Farrell and Crouchley, 1983) so that it may be assumed that the survivor data is a reasonable proxy of new firm formation. It is apparent that the rate of indigenous new firm formation rose sharply in 1977 (Table 5.1) partly because of the extension of the SI scheme to Dublin but higher rates also occurred throughout the rest of the urban system (Table 5.2). Buoyancy in terms of more new indigenous firms emerged somewhat later in Cork and the major provincial

Table 5.1 *Annual number of new indigenous single plant firms 1973–80 (inclusive)*

	1973	1974	1975	1976	1977	1978	1979	1980	Total
Total openings[1]	153	127	123	178	215	226	198	162	1,382
Survivors[2]	142	109	119	169	238	273	247	185	1,482

Notes: 1. Dublin is excluded; total incorporates survivors and new entries which subsequently closed.
2. Dublin is included.

Table 5.2 *Annual number of new indigenous single plant firms[1] by town size group, 1973–81*

Year	<1500	1,501–5,000	5,001–10,000	10,001–25,000	25,001–100,000	Cork	Dublin	Total
1973	39	30	16	11	6	12	28	142
1974	36	25	13	8	4	5	18	109
1975	34	26	18	11	6	4	20	119
1976	62	27	30	24	5	1	20	169
1977	62	46	23	36	8	7	56	238
1978	82	56	34	23	15	4	59	273
1979	56	35	26	27	26	22	55	247
1980	47	37	21	26	20	11	23	185
Total	418	282	181	166	90	66	279	1,482
Manufacturing employment 1973	27,374	31,253	22,783	23,290	16,484	15,289	91,494	227,967
Number of new ISPs per 1,000 manufacturing employees (1973) per annum	1·91	1·13	0·99	0·89	0·68	0·54	0·38	0·81

Note: 1. Data is for survivors only which permits inclusion of Dublin.

towns between 25,000 and 100,000. Indeed, the recent upward trend in new firm formation is somewhat more pronounced in the larger towns than elsewhere. Between 1973–7 (inclusive) the 25,000–100,000 towns, Cork and Dublin generated 25.7% of all new indigenous firm survivors while over the 1978–80 period 33.3% of all new firms were established in these larger towns. This net shift towards the metropolitan areas is only *partly* accounted for by the extension of the Small Industry programme to Dublin in 1977. The success of the other large towns might reflect the increased spin-off of new enterprises following the considerable phase of IDA-sponsored overseas investment in places such as Limerick-Shannon, Galway and Waterford since the late 1960s. When the number of ISP survivors is expressed per 1,000 manufacturing employees (1973) per annum, there is a clear inverse relationship between the rate of indigenous new firm formation and town size. The highest rate occurs in the below 1,500 population category (1.91) and the rate falls steadily with increasing town size so that all size groups above 25,000 have rates below the national average. The Dublin rate is only one-fifth that of the smallest communities (1,500) and about one-half of that recorded by the major provincial towns between 25,001 and 100,000 population (Table 5.2). A high rate of indigenous new firm formation is predominantly a rural-small town phenomenon: only 29% of the survivors were located in towns of over 25,000 although they accounted for 54% of industrial employment in 1973.

SPATIAL AND SECTORAL VARIATIONS IN NEW FIRM FORMATION

When indigenous new firm formation rates are calculated for the non-Dublin area thereby permitting the inclusion of survivors *and* new openings which subsequently closed, the national rate is 1.27 which compares with the corresponding survivors-only figure of 1.10 (Table 5.3). The increasing rate of new firm formation observable from 1976–7 is most pronounced in the North West, West, Mid West, South East, Kildare, Meath and Wicklow and the Midlands. The increasing rate of new firm formation observable in the Mid West region since the late seventies may partially reflect the changed role of the Shannon Free Airport Development Company to develop indigenous industry in an intensive and innovative way not hitherto attempted in Ireland. This scheme has been operating for too short a period to enable an unequivocal conclusion to be drawn concerning its effectiveness in the Mid West relative to the less resource intensive IDA programme in the rest of the country. In Donegal and the North East, there is no strong post-1976 upward trend.

Table 5.3 *Annual number of new indigenous single plant firms by region[1], 1973–81*

Year	Donegal	North West	West	Mid West	South West	South East	Kildare Meath Wicklow	North East	Midlands	Total
1973	13	4	21	11	31	14	13	22	24	153
1974	6	2	9	19	25	13	23	12	18	127
1975	9	3	17	14	15	17	17	20	11	123
1976	19	9	28	14	18	18	25	31	16	178
1977	5	15	19	26	21	44	30	24	31	215
1978	12	13	20	28	13	29	60	25	26	226
1979	3	6	29	19	40	33	20	20	28	198
1980	9	11	8	21	32	25	23	23	10	162
Total	76	63	151	152	195	193	211	177	164	1,382
Manufacturing employment 1973	5,422	3,246	8,269	18,274	35,182	23,142	15,703	18,098	9,137	136,473
Number of new ISPs per 1,000 manufacturing employees (1973) per annum	1·75	2·43	2·28	1·04	0·69	1·04	1·68	1·22	2·24	1·266

Note: 1. Dublin is excluded; total includes survivors plus new establishments which subsequently closed.

The highest rate of indigenous new firm formation per 1,000 manufacturing employees (1973) per annum is in the North West (2.43) – almost double the national average – with the West and Midlands also registering very high rates. The North East, the South East and Mid West have very similar rates – all with a strong urban base and some tradition of manufacturing industry – but nevertheless with below average rates of indigenous new firm formation (Table 5.3). The lowest rate of domestic owned new business formation in manufacturing industry was in the South West region (0.69). The highest rates, therefore, are in the most rural and least industrialized regions of the country. An exception to this pattern is the relatively high rate of new firm foundation in Meath, Kildare and Wicklow, which may be partly due to the movement out of founders from Dublin, where rates of formation are extremely low, to take advantage of less congestion and cheaper sites.

A regional level of aggregation is not the most appropriate one at which to analyse new firm formation, a process which is usually highly localized (Gudgin, 1978). Some interesting patterns emerge, therefore, when the regional rates are disaggregated by county (Table 5.4). Some counties, notably Leitrim, Roscommon and Longford, had such a small manufacturing base in 1973 that it was necessary to aggregate them with a contiguous county to produce rates of new firm formation which might be more validly compared with other areas. The Roscommon-Longford area has by far the highest rate of new firm formation, 4.44 firms per thousand employees per annum. It is interesting to note that their neighbouring counties in the Midlands region have much lower rates: Westmeath (2.42), Offaly (1.34) and Laois (1.29). This may be partly explained by the fact that Roscommon and Longford are in the Designated Areas with regional development grant levels of up to 60%, while in the rest of the Midlands, which is non-designated, the grant ceiling is 45%. However, other factors, such as size and sectoral mix, upon which attention will be focused in the econometric analysis, may also be partly responsible for inter-county variations in new firm formation rates.

The highest rates of new firm formation occur in a block of country south of Donegal and north of a line from Galway to Drogheda in the North West and West regions together with Longford-Roscommon, Westmeath, Cavan and Meath – all with rates of new firm formation above 2.0 per 1,000 manufacturing employees (1973) per annum (Fig. 5.1). These are, with the exception of the commuting zone of Co. Meath and around Galway city, predominantly rural counties.[2] By

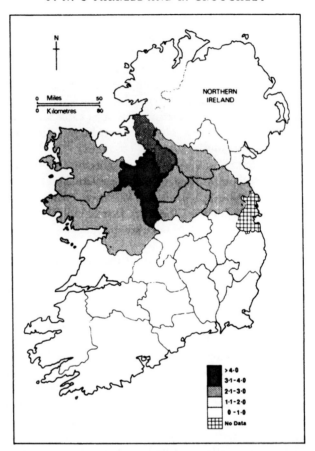

Fig. 5.1 Annual number of new firms per thousand manufacturing employees
(1973)

contrast, the counties of the South West, Mid-West and South East
have recorded low rates of indigenous new business formation in
manufacturing – less than one third of the rate achieved in the
Longford-Roscommon area. The lowest rates of all were recorded in
Dublin (where the survivor rate is about one-tenth of the Longford-
Roscommon figure), Cork, Tipperary N.R., Waterford, Tipperary
S.R. and Louth. Three of these are the most urbanized counties but the
low rates in Tipperary are seemingly anomalous.

There are very considerable between-sector differences in new firm
formation rates around the national average of 1.27 new firms per

Table 5.4 Number of new indigenous single plant firms by county, 1973–81

County	Total number of new ISPs	Number of ISP survivors	Manufacturing employment 1973	Number of new ISPs per annum per 1,000 manufacturing employees (1973)	Number of ISP survivors per annum per 1,000 manufacturing employees (1973)
Donegal	76	70	5,422	1·75	1·61
Leitrim	24	20	764	3·93	3·27
Sligo	39	39	2,482	1·96	1·96
Galway	99	86	5,182	2·39	2·07
Mayo	52	40	3,087	2·11	1·62
Clare	58	43	6,560	1·11	0·82
Limerick	77	72	8,461	1·14	1·06
Tipperary N.R.	17	15	3,253	0·65	0·58
Cork	156	142	30,407	0·64	0·58
Kerry	39	38	4,775	1·02	0·99
Carlow	25	22	2,562	1·22	1·07
Kilkenny	34	29	3,262	1·30	1·11
Tipperary S.R.	27	24	3,958	0·85	0·76
Waterford	54	48	8,847	0·76	0·68
Wexford	53	49	4,513	1·47	1·36
Dublin	–	279	91,494	–	0·38
Kildare	56	49	6,312	1·10	0·97
Wicklow	75	62	5,069	1·85	1·53
Meath	80	69	4,322	2·31	2·00
Cavan	47	40	2,796	2·10	1·79
Louth	82	65	11,790	0·87	0·69
Monaghan	48	39	3,512	1·71	1·39
Laois	20	20	1,934	1·29	1·29
Longford	23	17	1,045	2·75[1]	2·03[2]
Offaly	35	31	3,264	1·34	1·19
Roscommon	49	42	982	6·24[1]	5·35[2]
Westmeath	37	32	1,912	2·42	2·09
Total	1,382	1,482	227,967	1·27	0·81

Notes: 1. Pooling the contiguous counties of Longford and Roscommon gives a rate of 4·44.
2. Pooling Longford and Roscommon produces a rate of 3·64.

Table 5.5 *Number of new indigenous single plant firms[1] by sector, 1973–81*

Sector[2]	Total number of new ISPs	Number of new ISP survivors	Manufacturing employment 1973	Number of new ISPs per annum per 1,000 manufacturing employees (1973)	Number of new ISP survivors per annum per 1,000 manufacturing employees (1973)
1. Bacon and slaughtering (4,5)	13	12	8,027	0·202	0·187
2. Creamery butter and milk products (6)	5	4	7,001	0·089	0·071
3. Grain milling and animal feed (8)	12	9	4,274	0·351	0·263
4. Bread, biscuits (9)	24	23	4,884	0·614	0·589
5. Jam, canned food, sugar, cocoa, chocolate, margarine, miscellaneous food (7,10,11,12,13)	51	41	6,525	0·977	0·785
6. Drink and tobacco (14–18)	7	6	4,930	0·178	0·152
7. Woollen and worsted, linen, cotton, jute, nylon etc. (19–21)	30	23	8,411	0·446	0·342
8. Hosiery (22)	14	8	5,317	0·329	0·188
9. Boot and shoe (23)	7	4	4,911	0·178	0·102
10. Clothing (24)	68	54	9,183	0·926	0·735
11. Made-up textiles (25)	17	15	4,815	0·441	0·389
12. Wood, cork, brushes (26,28)	78	72	3,056	3·190	2·945
13. Furniture (27)	214	197	5,813	4·602	4·236
14. Paper/paper products (29)	13	12	1,348	1·205	1·113
15. Printing and publishing (30)	42	34	3,334	1·575	1·275

16. Miscellaneous including fell-mongery and leather (31,32,47)	85	68	9,516	1·117	0·893
17. Fertilizers, paints, chemicals, soap, pharmaceuticals (33–36)	33	29	4,069	1·014	0·891
18. Glass, pottery (37)	34	30	4,138	1·027	0·906
19. Cement, structural clay (38,39)	70	63	7,410	1·181	1·063
20. Metal trades (40)	359	330	12,490	3·593	3·302
21. Machinery manufacture (41)	42	34	3,290	1·596	1·292
22. Electrical machinery (42)	62	51	5,315	1·458	1·199
23. Shipbuilding and repairs, railroad equipment, road and other vehicles (43–46)	38	33	4,383	1·084	0·941
24. Plastics (48)	41	33	1,970	2·602	2·094
25. Construction, agriculture, services (60,65,70)	23	18	2,062	1·394	1·091
Total	1,382	1,203	136,473	1·266	1·101

Notes: 1. Table excludes Dublin; there were 279 new ISPs in Dublin which survived until 1981. The national rate of new firm formation for survivors including Dublin is 0·813 compared with 1·101 for the country outside Dublin.
2. Numbers in parentheses are IDA Product Codes.

annum per 1,000 manufacturing employees (Table 5.5). The highest
entry rate occurs in furniture (4.6) followed by metal trades, wood,
cork and brushes, and plastics – all industries with relatively low entry
barriers. The only other sectors with entry rates above the national
average are printing and publishing, machinery manufacture, electrical
machinery and construction, agriculture and services (Table 5.5). The
lowest rates of entry were recorded by butter and milk products (0.09),
drink and tobacco, boots and shoes, bacon and slaughtering, and
hosiery. These figures demonstrate that between industry variation in
new firm formation is very high: the rate of entry in furniture, for
example, is twenty-five times greater than in either drink and tobacco
or boots and shoes. Exit rates expressed as the number of the 1973 stock
of plants closing per 1,000 employees (1973) range from 1.21 (wood,
cork) and 1.65 (furniture) to bacon and slaughtering (0.16) and ferti-
lizers, chemicals and pharmaceuticals (0.17). It appears that industries
experiencing high exit rates also record high rates of entry. The
turnover rate of plants – i.e. entry and exit rates – is responsible for the
age distribution of each industry and it has been shown elsewhere
(O'Farrell and Crouchley, 1983) that age of young plants is related to
closure (exits). Hence, sectors with high entry rates, *ceteris paribus*, will
have high exit rates both because of a higher proportion of young plants
and as a consequence of size since most young plants are small and therefore more vulnerable to closure. As expected, entry rates and exit rates
of new firms are quite highly correlated – $r^2 = 0.67$ for all entries –
although the relationship is purely associative.

New firm formation rates: Irish and UK comparisons

The comparisons of new firm formation rates in manufacturing
between Ireland and the East Midlands of England need to be interpreted with great caution and small differences should not be regarded
as in any way meaningful for a number of reasons. First, comparisons
are hampered by differences in data sources and survey methodologies.
Second, the industrial classification systems are not identical although
every effort has been made to include only those groups which appear
to be very similar by aggregating up from MLH level. Third, the time
periods over which the annual rates are calculated are different in the
two areas (Table 5.6). Examination of the Irish and East Midlands
industry rates shows that the number of new firms per 1,000 manufacturing employees in the base year per annum in Ireland, 1.01, is
considerably higher than the corresponding East Midlands figure of

Table 5.6 *Indigenous new firm formation rates[3] by
industry: UK and Irish comparisons*

Industry	East Midlands, England 1968–75	Ireland 1973–81
Food	0·39	0·36
Drink and tobacco	0·07	0·15
Hosiery	0·36	0·19
Footwear	0·14	0·10
Clothing	0·75	0·74
Wood, cork, brushes	2·52	2·95
Furniture	1·95	4·24
Paper products	0·43	1·11
Printing and publishing	0·79	1·28
Chemicals, fertilizers, pharma- ceuticals, etc.	0·27	0·89
Cement, structural clay, glass and pottery	0·27	1·01
Plastics	1·85	2·09
Total	0·42	1·01

Notes: Formation rate: number of new indigenous firms per 1,000 manufacturing employees in the base year per annum (data represents survivors only).
Sources: Ireland (this study)
East Midlands (Steve Fothergill, University of Cambridge, kindly supplied the data at MLH level).

0.42 (Table 5.6). In addition, employment generated by new firms in Ireland between 1973–81 as a percentage of base year manufacturing employment (7.9%) was substantially greater than the East Midlands 1968–75 figure of 4.1%. Specific industry comparisons reveal that the rate of new firm formation is much higher in paper products, and furniture in Ireland as it also is in printing and publishing, chemicals and pharmaceuticals, cement, structural clay and glass (Table 5.6). The East Midlands recorded higher rates in hosiery and footwear – industries in which there are major concentrations in this English region – and several industries, notably food, clothing, wood, cork and brushes registered very similar new firm formation levels in the two areas.

Furthermore, the number of new indigenous firms per annum in Ireland per 1,000 manufacturing employees in the *end year* is 6.20 which is considerably higher than the rates for South Hampshire (3.2), the East Midlands (3.0), Cleveland (1.3) or Scotland (1.0) (Table 5.7). Similarly, the impact of new firm survivors upon *end year* employment

Table 5.7 *International comparisons of new firm entry rates*

	Entry rate per annum[1]	Employment in new firms as % of total end year employment	New firms per 1,000 manufacturing employees in end year
USA (all USA[a])	1950–58	6·9		
Canada (Ontario[b])	1961–65	6·1		
Norway[c]	1937–48	5·0		
UK: Leicestershire[d]	1947–55	3·1		
South Hampshire[e]	1971–79		3·5	3·20
Cleveland[f]	1965–76		1·8	1·30
East Midlands[g]	1968–75		4·2	3·0
Scotland[h]	1968–77		2·2	1·0
Ireland[i]	1973–81	5·7	7·5	6·20

Note: 1. Entry rate defined as number of new indigenous firms per annum as a percentage of total population of establishments at beginning of period (see text for departures from this definition in some studies).

Sources: a. CHURCHILL, 1959
   b. COLLINS, 1972
   c. WEDERVANG, 1965
   d. GUDGIN, 1978
   e. MASON, 1982
   f. ROBINSON and STOREY, 1981
   g. FOTHERGILL and GUDGIN, 1979
   h. CROSS, 1981
   i. This study

in Ireland is approximately double that for either the East Midlands or South Hampshire in the UK (Table 5.7).

New firm formation rates: international comparisons

It is not possible to use the measure of entry rate expressed per 1,000 industrial employees for the purposes of international comparisons because most researchers have defined entry rates per annum as a percentage of the number of firms (or establishments) in the base year. The measure used throughout this study is preferable since it reflects the process by which the population of industrial employees is the relevant indicator of the number of potential entrepreneurs (Gudgin, 1978, p. 136). The base measured in terms of number of firms (or plants) fails to take account of the size of the latter. However, in order to make any international comparisons, this less satisfactory measure must be used. The national studies for the USA (Churchill, 1959) and Norway (Wedervang, 1965) refer to firms while the UK and Canadian data relate to establishments and are, therefore, more directly comparable with those for Ireland. Furthermore, the US data classified changes of ownership as births and deaths so great care must be exercised in comparing the findings. The results presented in Table 5.7 suggest that entry rates in Ireland are comparable with those for Norway (between 1937–48), Canada and the USA (especially when allowances are made for changes in ownership) but the UK figure is over 40% lower than the Irish one. When the employment impact of new firm survivors is considered relative to total end year employment, the Irish figure of 7.5% is approximately double those for the East Midlands and South Hampshire in the UK (Table 5.7).

SECTORAL VARIATIONS IN NEW FIRM FORMATION RATES: SOME THEORETICAL CONSIDERATIONS

In analysing entry rates of new firms to various industries, the conceptual framework will be derived primarily from economics since this paper is concerned with *aggregate* rates of new firm formation and not the micro-process of formation at the individual level. At the micro level, social psychologists and sociologists have developed insights into those factors which motivate a person to set up his own firm but such factors may be assumed to be randomly distributed at the aggregate inter-industry level.

The traditional economic view is to conceive of the entry of new firms into an industry as a reflection of the chances of profit making in that industry.[3] Formation rates are assumed to be a function of perceived post-entry profitability and the real or expected entry barriers so that economists have tended to ignore the supply side of entrepreneurship (Storey, 1982, p. 3). Under the unrealistic assumptions of the static perfect competition model, new firms enter markets where prices persistently exceed long run average cost.[4] In imperfect markets, however, where pricing policy may be influenced by the entry of new firms, rates of formation may be depressed by barriers to entry. Bain, 1956, argued convincingly that sellers in certain industries can raise prices above a competitive level without attracting new firms into the industry. There are a wide range of such barriers including: product differentiation; control over input suppliers and/or outlets; scale economies; legal and institutional factors including patents; large capital requirements; degree of seller concentration; absolute costs disadvantages and so on.

Contemporary theory of the firm, including most of the managerial models, does not incorporate any distinctive entrepreneurial function (Johnson and Darnell, 1976, p. 5). There is no consensus upon the precise nature of entrepreneurship, except that it involves something more than the daily management of the firm. Included in the activities attributed to the entrepreneur have been: risk taking; combination and organization of the factors of production; leadership and motivation; long range planning; and, especially, innovation. Entrepreneurship, therefore, can and does occur within an *existing* business. Conversely, entrepreneurship and new firm formation are *not* synonymous; there are many types of new enterprise – most of which display few, if any, entrepreneurial characteristics. Schumpeter (1961, p. 66), however, saw a clear link between the entrepreneur (as he defined it) and the new firm, although he conceded that there were exceptions. Many, probably most, new firm founders duplicate existing production functions and orientate towards existing markets and supply sources. The new firm founder is more likely to be an adapter and imitator than an innovator, to have more in common with Marshall's organizer of factors of production than Schumpeter's creator of new products or processes.

A number of researchers (Mansfield, 1962; Gudgin, 1978; and Johnson and Cathcart, 1979) have analysed, within a multiple regression framework, inter-industry difference in formation rates. In this study, the dependent variable Y, is defined as the number of new indigenous single plant firms per annum per 1,000 employees in industry *i* in the base year entering between January 1973 and January 1981. The independent variables have usually included: (1) the proportion of plants in industry *i* employing less than twenty in the base year (X_1); and (2) employment change in industry *i* over the study period (X_2). These variables are usually specified to reflect, respectively, the extent to which barriers to entry exist in any given industry and the attractiveness of industry *i* to potential entrants.[5] Employment in a small firm is assumed to be a better preparation for founding a business because of the wide range of task experience derived, the opportunity of regular contact with the director (who may also be the founder) and the lower level of salaries, fringe benefits and job security than in larger firms. Hence, sectors containing a high proportion of small firms will, *ceteris paribus*, generate more new founders. However, the proportion of small plants in an industry does not necessarily simply reflect entrepreneurial supply but may also mirror ease of entry (i.e. low capital requirements) or the youthfulness of an industry (Cross, 1981, p. 168). It is hypothesized that the relationship between variable X_1 (percentage of plants in sector employing fewer than twenty persons in 1973) and the dependent variable (Y_1) is positive. It is not possible to disentangle all the micro-processes underlying variable X_1 within an aggregate model. The regression parameter will determine whether variable X_1 is significant; further work at the level of individual founders and their firms will be necessary to identify the mechanisms underlying it.

The extent to which new firm founders move into a different industry product category from their previous employment represents, together with the introduction of new products and processes by firms already in the region, an important element in the process of diversifying regional economies. Variable X_2, the employment change of the 1973 stock of plants in industry *i*, is tested against the rate of entry to verify the hypothesis that sectors experiencing growth will attract more new firms. Since a person working in industry *i* is more likely to identify a market gap in industry *i* than an individual working in *j*, any analysis of inter-industry formation rates must take account not only of the relative attractiveness of different industries as a *destination* for new formations (the growth rate variable), but also the relative suitability of such industries as generators of spin-off (Johnson and Cathcart, 1979, p. 278). The regression results may be picking up not only the attractiveness of a given industry as a destination for new founders but also its 'effectiveness' as a source industry. There is no way within this analytical framework that the two effects can be isolated. Johnson and

Cathcart, 1979; and Gudgin, 1978, appear to have measured employment change in all plants so that their expansion rates are dependent, at
least in part, upon the rates of entry into each industry. In the absence of
any strong theoretical arguments in favour of a specific functional
form, a linear relationship is usually fitted (see Johnson and Cathcart,
1979).

$$Yi = a + b_1 X_1 + b_2 X_2 + e$$

Both Gudgin, 1978, and Johnson and Cathcart, 1979, showed that
variable X_1, was significant but not X_2 – the employment change
factor.

Plant dominance and industrial concentration are often viewed as
factors which act to inhibit the rate of entry (Cross, 1981, p. 167). First,
the dominance of an industrial sector by a few large plants would
suggest that the minimum size for efficient production is large and
would, therefore, discourage entry. A second influence of large plants
refers to the supply side of founders: the structure and work experience
gained while working for a large organization is not conducive to entrepreneurship. A number of workers have observed that spin-off rates
from smaller plants are much greater than from larger ones (Johnson
and Cathcart, 1979; Cooper, 1971). This variable (X_3) is specified as the
percentage of employment in industry i located in plants employing
over 200 persons. The variable ranges from 0 (plastics) to 79.8% (electrical machinery) and is conceived as inhibiting new firm entry. The
large plant size effect may be *reinforced* by the *ownership* factor (X_6).
Increasing external ownership may decrease the number of risk taking
managerial positions which reduces the potential supply of founders.
However, this variable may also reflect demand side factors such as the
greater opportunities for MNEs to achieve scale economies. It is opera-
tionalized as the proportion of employment in the sector controlled by
multiplant firms (MNEs and IMPs). Cross, 1981, p. 172, reported that
the level of Scottish ownership was positively related to sectoral variations in entry rates although this result should be interpreted with some
caution since it is based upon rank correlations without control for other
factors. Variable X_4 – the median size of new ISPs in each sector at end of
first year – is included in order to test the hypothesis that the smaller the
initial size at which a firm can become established in an industry, the
easier it will be for new founders to enter. This may be regarded as a
crude proxy of the capital requirements for an operation of minimum
efficient size (for definition of variables see Appendix 2).

It is also hypothesized that there is an inverse relationship between median age of the 1973 stock of plants in an industry (X_5) and the rate of entry. The inclusion of age may reflect a number of processes: (1) a high turnover rate within an industry; (2) a youthful new industry (e.g. electronics) with a low average age; and (3) a direct relationship between the youthfulness of an industry and its ability to produce new firms (Cross, 1981, p. 172). The causal mechanism is ambiguous. A host of personal and micro-environmental factors may also be important but it is expected that personality and attitudinal factors to entrepreneurship will operate randomly at an aggregate inter-industry level. Thus, the econometric analysis will permit examination of the aggregate patterns of entry and identification of some of the major influences upon it but cannot differentiate between the micro-level variables. The inter-industry analysis of entry will be conducted upon all firms[6] which entered between January 1973 and January 1981 (Y_1).

SECTORAL VARIATIONS IN RATES OF NEW FIRM FORMATION: AN ANALYSIS

The twenty-five industry groups outlined in Table 5.5 were selected in consultation with the IDA in order to maximize product homogeneity and minimize employment size variance between industries. The latter constraint, necessary for regression analysis, prevented a more detailed disaggregation of industry groups. *A priori* theoretical reasoning and previous empirical evidence determine the order of entry of the independent variables. The value of X_1 (the percentage of plants employing fewer than twenty persons) ranges from 77% in furniture to 20% in boots and shoes; while X_2 (percentage employment change 1973–81 of the 1973 plant stock) ranges from a decline of 51% in woollen, worsted, linen and cotton to an increase of 95.5% in bacon and slaughtering. Some seventeen sectors declined and only eight expanded employment. Table 5.8 shows that X_1 (the proportion of plants employing fewer than twenty) is positively related, as hypothesized, to entry rates thereby confirming the results obtained by Wedervang, 1965, pp. 178–9; Johnson and Cathcart, 1979, p. 277; and Gudgin, 1978, p. 166. Equation 2 is specified to reflect both the effect of X_1 and the employment change variable, X_2; the results show that X_2 is insignificant which supports the findings of other studies. Similarly, although the sign of the co-efficient X_3 (proportion of employment in establishments employing over 200) is negative, as postulated, the variable is not significant; its inclusion has a negligible effect upon the

Table 5.8 *Regression equations with the number of new indigenous single plant firms per 1,000 employees in 1973 per annum for 25 industry groups as dependent variable (Y_1)[1]*

Equation	Equation number	$\bar{R}^2$	S.E.E.[2]
$Y_1 = -0.89 + 0.04\ X_1$ $\qquad\qquad (3.04)^{**}$	(1)	0.131	0.926
$Y_1 = -0.93 + 0.04\ X_1 - 0.005\ X_2$ $\qquad\qquad (2.52)^{*}\qquad (0.78)$	(2)	0.32	0.916
$Y_1 = -0.02 + 0.03\ X_1 - 0.01\ X_3$ $\qquad\qquad (2.52)^{*}\qquad (1.52)$	(3)	0.41	0.793
$Y_1 = 0.16 + 0.03\ X_1 - 0.06\ X_4$ $\qquad\qquad (1.64)\qquad (0.78)$	(4)	0.32	0.926
$Y_1 = 0.42 + 0.05\ X_1 - 0.06\ X_5$ $\qquad\qquad (3.93)^{**}\quad (3.05)^{**}$	(5)	0.65	0.475
$Y_1 = 1.38 + 0.04\ X_1 - 0.07\ X_5 - 0.02\ X_6$ $\qquad\qquad (3.20)^{**}\quad (3.94)^{**}\quad (2.33)^{*}$	(6)	0.76	0.311

Notes: 1. Y_1 includes survivors and new firms which opened and subsequently closed.

2. S.E.E. = standard error of the estimate.

*Significant at $p < 0.05$.   **Significant at $p < 0.01$

For each equation the figures in parentheses are t values.

parameter and standard error of X_1 (Table 5.8). In the case of variable X_4 (the size of new ISPs at entry) the direction of the relationship is negative, as predicted, but the variable is not significant. The final series of hypotheses examines the rate of entry and its relationship with several characteristics of each industry. Variable X_5 (the median age of the 1973 stock) ranges in value from 4.5 years (plastics) to 35 (creamery, butter and milk; and drink and tobacco). The relationship between X_5 and Y_1 is negative, as postulated, and is significant $(p < 0.01)$.

It is also hypothesized that there is a negative relationship between the percentage of employment in each industry controlled by multiplant organizations and the rate of new firm formation. The highest proportions of employment controlled by multiplants occur in paper and printing (88.4%), machinery manufacture (84.0%) and miscellaneous (82.9%) with the lowest in furniture (15.7%), printing and publishing (15.9%) and bread and biscuits (25.8%). When variable X_6 is entered into an equation containing X_1 and X_5, it is significant with a negative directional relationship, as predicted (equation (6) Table 5.8). Hence, this is the best fit model with an $\bar{R}^2 = 0.76$ and the direction of all relationships as originally hypothesized.

The parameters of equation (6) Table 5.8 suggests that a one percentage point increase in the percentage of plants employing fewer than twenty people increases the rate of new firm formation by 0.04 per 1,000 employees per annum, having allowed for the effects of age and proportion of employment in multiplant firms.[7] This parameter is higher than the one of 0.02 reported by Gudgin, 1978, p. 137, in the case of the East Midlands of Britain thereby implying a stronger size effect in Ireland. The coefficient of X_5, median age, suggests that a 1% rise in age of the manufacturing plant stock is associated with a lowering of new firm formation rates by 0.07 per 1,000 manufacturing employees per annum, although a causal relationship is not implied since age may reflect a number of processes. Finally, a one percentage point rise in the percentage of employment owned by multiplant enterprises reduces new firm formation rates by 0.02. An inspection of standardized residuals from equation (6) of Table 5.8 suggests that those sectors where indigenous new firm formation rates have been higher than those predicted by the regression equation are furniture, metal trades, plastics, and wood, cork and brushes. These are all industries characterized by very high rates of new firm formation implying that there are other factors, in addition to those included in the model, which explain their performance. One such factor is that most new firms in these industries produce products with a high bulk or weight to value ratio which for logistic reasons are subjected to little or no competition from imports (i.e. the new enterprises are relatively sheltered). Sectors with the largest negative residuals where new firm formation rates are *lower* than those predicted by the model are shipbuilding and repairs, railroad equipment and vehicles, bread and biscuits, printing and publishing, and cement and structural clay. These are clearly sectors which are more difficult to enter than predicted by the variables in the model; there are factors other than plant size, age and extent of multiplant ownership influencing rates of new firm formation in these sectors. These might include initial capital requirements, price discrimination, state of market demand, scale economies, degree of foreign competition and so on. Since the causal mechanisms underlying the statistical associations in the model are complex and not fully understood and since the number of observations is small (twenty-five), the models are of limited value for the purpose of policy prescription.

The analysis of residuals has suggested that much of the growth in new firm formation has been in business relatively sheltered from international competition.[8] It is not possible to quantify this factor precisely and incorporate it in the econometric model. However, it is apparent that one-third of new firms are in predominantly non-traded sectors of paper, printing, packaging, wood, furniture, cement, glass and clay (Table 5.5). Only 10% of new firms have been established in overwhelmingly traded (i.e. open to international competition) sectors such as clothing, footwear and textiles while the remainder (57%) are attributable to heterogeneous industries like metals and engineering, plastics, food and consumer goods (National Economic and Social Council, 1982, pp. 15–16).

The overall conclusion of this exercise is that new firm entry rates differ substantially between sectors; that entry rates are quite strongly related to the size structure of an industry but are not correlated with employment change; that the age of the existing plant stock and the extent to which the employment in an industry is controlled by multiplant organizations are also related to entry rates; and that the highest formation rates have occurred in non-traded businesses largely protected from international competition. Although all of these statistical associations with rates of entry are plausible, caution must be urged in making casual inferences.

SPATIAL VARIATION IN RATES OF NEW FIRM FORMATION:
A REGRESSION ANALYSIS

An analysis of spatial variations in entry rates should help identify some of those characteristics of local economies which stimulate or inhibit the rate of new firm births. Entry rate variations have already been reported at international, inter-regional and inter-county level. Secondary data is available at county level in Ireland and so this is the smallest scale at which an analysis can be conducted. The dependent variable (Y_3) is defined as the number of new indigenous single plant firms per annum per 1,000 employees in manufacturing industry in county j in the base year (1973). Variable (Y_3) includes survivors and those firms which opened and subsequently closed.[9] Dublin is not included in the model because the size distribution of manufacturing employment in the country is not known for 1973 and, therefore, variable X_7 (proportion of total manufacturing plants employing less than twenty in 1973) could not be calculated for the Dublin area.

Some of the independent variables specified in the spatial analysis are identical or similar to those incorporated into the inter-industry model and the reader is referred to the earlier theoretical section for a discussion of these factors. It is hypothesized, as in the inter-industry model, that the percentage of manufacturing plants in county j employing fewer than twenty persons in 1973 will be positively related to the rate of new firm formation. However, although entry rates may be a function of the proportion of small plants in an area, a higher percentage of small plants may be the *result* of higher entry rates in the past. The population density of an area may directly influence the size of plants. Low population density areas attract relatively fewer large plants partly because a small town or village catchment cannot provide the quantity of labour required. Hence, less urbanized counties will possess a higher proportion of small plants. The size factor is a catch-all for several plausible influences and, therefore, care must be exercised in interpreting a significant size coefficient.

It is postulated that there is a positive relationship between the rate of change in manufacturing employment in county j, 1973–81 (X_8) and the rate of new firm formation. The expansion of a county's manufacturing base may open up new markets and increase existing ones thereby providing opportunities for new firms. Rising manufacturing employment will also increase the pool from which new founders are most likely to emerge (Cross, 1981, p. 263).

There is evidence that large plants are poor incubators of new firm founders compared with small establishments (Johnson and Cathcart, 1979). Hence, it is suggested that the greater the extent to which manufacturing employment is concentrated into large plants employing over 200 persons (X_8), the lower will be the rate of indigenous new firm formation. A relatively large agricultural sector might enhance the new firm formation rate in manufacturing since farmers have direct experience of self-employment and decline of agricultural employment is continuing throughout the country thereby adding to the supply of potential entrepreneurs. It is postulated that the percentage of employment in agriculture in county j in 1971 (X_{10}) is positively related to the rate of new firm formation.

Outside the manufacturing sector itself, the greatest potential pool of new founders probably exists among the commercial and retailing business community, many of whom are self-employed. It is suggested that the percentage of employment in commerce, retailing and wholesaling in county j in 1971 (X_{11}) is positively related to the rate of new firm formation and that X_{11} may be conceived as a crude proxy for the level of non-manufacturing entrepreneurship in an area.

Gudgin, 1978, reported higher rates of new firm formation in rural areas of the East Midlands than in the older industrial towns, a phenomenon which he explained by rural districts adjacent to towns acting as overspill and by long distance commuters dwelling in rural areas preferring a location close to their homes. The form of the relationship may not be the same in Ireland where towns throughout the hierarchy have attracted significant industrial development during the past twenty years. The rural factor is investigated here by specification of an interval level variable, X_{12} (proportion of a county's population in towns of over 5,000 population), to test the hypothesis that rates of indigenous new firm formation are lower in more urbanized counties. A number of processes, in addition to those identified by Gudgin, 1978, such as lack of alternative job opportunities, may underlie the rural–urban differential in indigenous new firm formation.

Johnson and Cathcart, 1979, p. 278, also demonstrated that immigrant mobile plants in the Northern Region of England were relatively poor incubators of founders of new businesses. Hence, it is postulated that areas with a high proportion of their manufacturing employment within ISPs (variable X_{13}) will generate more new manufacturing enterprises. The median age of the 1973 stock of plants in county j (X_{14}) is used to investigate the proposition that industrial areas with a young manufacturing stock will have a higher rate of entry by new indigenous enterprises than areas with a mature manufacturing stock. The hypothesis is relatively simple but again there are problems of inference. Do more recently established plants have the highest spin-off rates? A young age structure might reflect the general growth of the existing level of manufacturing activity in the area and, therefore, could be a proxy measure for general manufacturing growth (Cross, 1981, p. 267). This factor may be controlled, however, through variable X_8.

The order in which the variables were introduced into the model was decided prior to the model fitting by *a priori* theoretical considerations and upon the basis of evidence from other research. The proportion of county manufacturing employment in plants employing below twenty in 1973 (X_7) ranges from 82% in Leitrim and 75.7% in Mayo to a minimum of 39.8% in Kildare. Variable X_7 is significant with the direction of the relationship being positive, as hypothesized (Table 5.9). The highest rate of manufacturing employment change between 1973–81 occurred in Longford (+130%) and Roscommon (+118%) with the lowest in Louth (−13.6%) and Monaghan (+2.4%). When this variable X_8 is fitted X_7 becomes insignificant in both models with the sign of X_8 being positive as postulated. In a model specified to include X_8 alone, the degree of correlation is quite high ($\bar{R}^2 = 0.65$); no other variables are significant when added to X_8 (equation (3) Table 5.9). The relationship suggests that for every one percentage point rise in manufacturing employment in the county, the entry rate of indigenous new firms rises by 0.02 per 1,000 employees per annum.

The value of variable X_9, the percentage of county manufacturing employment in plants employing over 200 in 1973 ranges from 62.3% in Waterford and 61.1% in Louth to zero in Leitrim, Laois and Longford. The direction of the relationship between X_9 and the dependent is negative, as suggested, but is not significant. The proportion of the workforce employed in the agricultural sector attains a maximum in Roscommon (60.3%) and Leitrim (59.5%) and minimum in Louth (12.6%) and Waterford (22.1%). When X_{10} is added to a model containing X_8, it does not contribute to increasing the proportion of variance explained. The counties with the highest proportions employed in commerce, retail and wholesaling are Wicklow (21.2%) and Limerick (19.9%) with Leitrim (9.8%) and Roscommon (11.5%) registering the lowest percentages; indeed the striking feature of variable X_{11} is its low variance between counties. The negative relationship between X_{11} and the dependent variable is not as predicted but is not statistically significant at $p < 0.05$.

The degree of urbanization in 1976 as expressed by the percentage of

Table 5.9 Regression equations with the number of new indigenous single plant firms per 1,000 manufacturing employees in 1973 per annum by county as dependent variable $(Y_3)^1$

Equation	Equation number	$\bar{R}^2$	S.E.E.^{2}
$Y_3 = -1.73 + 0.06\, X_7$ (3.29)**	(1)	0.31	1.00
$Y_3 = -0.47 + 0.03\, X_7 - 0.02\, X_8$ (1.43)* (3.49)**	(2)	0.69	0.45
$Y_3 = 0.87 + 0.02\, X_8$ (5.00)***	(3)	0.65	0.51
$Y_3 = -0.40 + 0.05\, X_7 - 0.02\, X_9$ (2.39)* (1.64)	(4)	0.42	0.85
$Y_3 = -1.80 + 0.03\, X_7 - 0.05\, X_{10}$ (2.48)* (1.48)	(5)	0.55	0.66
$Y_3 = 3.12 + 0.03\, X_7 - 0.19\, X_{11}$ (2.28)* (1.28)	(6)	0.49	0.74
$Y_3 = -0.03 + 0.04\, X_7 - 0.03\, X_{12}$ (2.20)* (2.17)*	(7)	0.50	0.72
$Y_3 = -1.82 + 0.05\, X_7 - 0.01\, X_{13}$ (2.43)* (0.95)	(8)	0.33	0.97
$Y_3 = -0.80 + 0.06\, X_7 - 0.07\, X_{14}$ (3.19)** (1.76)	(9)	0.39	0.88

Notes: 1. Y_3 includes survivors and new firms which opened and subsequently closed.
2. S.E.E. = standard error of the estimate.
*Significant at $p < 0.05$. **Significant at $p < 0.01$ ***Significant at $p < 0.001$.
For each equation the figures in parentheses are t values.

a county's population resident in towns of over 5,000 population ranges from a peak of 58.6% in Louth and 44.9% (Limerick) to zero in Leitrim, Roscommon, Longford and Cavan. The relationship between the degree of urbanization and the rate of new firm formation is significant and negative as expected (Table 5.9).

The proportion of manufacturing employment in ISPs in 1973 ranged from a maximum of 71.3% in Leitrim and 61.6% in North Tipperary to a very low minimum of 15.8% in Clare and 18.0% in Sligo. The relationship between X_{13} and the dependent variable is positive as predicted but is not significant. Similarly, X_{14} (the median age of the manufacturing plant stock in 1973) ranging from 26 years in North Tipperary to 4 in Leitrim, has the expected negative sign but is not significant.

The preferred model specification is that containing variables X_7 and X_{12} and not the equation with X_8 as the sole independent variable. Hence, equation (7) of Table 5.9 is favoured over equation (3) on grounds of *a priori* theoretical criteria and previous empirical evidence despite the fact that the latter equation has a higher $\bar{R}^2$ value; statistical criteria do not take precedence over theoretical ones. The parameters of equation (7) Table 5.9 suggest that a one percentage point decrease in the population of a county's population resident in towns over 5,000 population increases the rate of new firm formation by 0.03 per 1,000 employees per annum, having controlled for the effects of plant size. The same equation implies that a one percentage point increase in the proportion of plants in a county employing fewer than twenty boosts new firm formation by 0.04 per 1,000 employees per annum.

An examination of standardized residuals from equation (7) of Table 5.9 indicates that the counties where new formation rates are *higher* than those predicted by the regression equation including both X_7 and X_{12} are Roscommon, Sligo, Leitrim, Limerick and Louth. These are counties with some of the lowest and highest urbanization rates. Counties with the largest negative residuals where new firm formation rates are *lower* than those predicted by the model are Offaly, Laois, Kilkenny, North Tipperary, Kerry, and South Tipperary. These, with the exception of Kerry, are a block of central and south midland counties in the non-designated areas where, according to the model, higher levels of new firm formation should be expected. The contrasts between counties in the rate of new firm formation are quite large: the rates in Cork and Tipperary North Riding (0.65 and 0.64, respectively) are some seven times lower than that of Roscommon/Longford (4.44). The fact that less urbanized areas have higher rates of new firm formation confirms the results of both Gudgin, 1978; Cross, 1981; and Mason, 1982. There is a need, however, to unravel the causal factors underlying the rural/urban variable. Are higher rates in rural areas, having controlled for the plant size distribution, due to a more favourable sectoral mix, cheaper land prices, the wider availability of suitable premises, the lack of many alternative employment opportunities or some other process? Subsequent research may reveal the nature of the processes responsible for higher rates of business formation in rural areas. Clearly, sectoral mix influences entry rates at county level but it was not possible to include this factor in the model since many sectors have few or no employees in some counties and therefore meaningful comparisons of inter-county variation in number of new firms per 1,000 employees cannot be drawn. However, counties with low barriers to entry sectors – such as timber and furniture – will have higher entry rates as a consequence of this compositional effect. It must be acknowledged that aggregative regression analysis identifies some of the important factors in the process of new firm formation that are amenable to measurement but will not isolate all proximate causes. New firm formation is a complex process and many factors will only emerge through in-depth investigation at micro level. Some of the inter-county variation, for example, *may* be due to the varying performance of County Development Officers liaising between the new firm founders and the IDA.

CONCLUSIONS AND IMPLICATIONS

New firm formation is important for economic development and regional differences in entrepreneurship may be a partial explanation of variations in regional economic performance. Hence, an important aim of regional policy should be to stimulate growth in locally owned plants, especially through the encouragement of new firm formation.

An analysis of new firm formation rates at the level of twenty-five industry groups indicates that between industry variation in entry is high ranging from a maximum in furniture, followed by metal trades and wood, cork and brushes with the lowest rates recorded by butter and milk products, drink and tobacco, and boots and shoes. The regression analysis indicates that there are three variables associated with the dependent, number of new indigenous firms per annum per 1,000 employees in sector i (1973): (1) the proportion of plants in the sector employing fewer than twenty people; (2) the median age of the sector's plant stock in 1973; and (3) the percentage of sectoral employment controlled by multiplant organizations in 1973. Furthermore, there is some evidence to suggest that formation rates have been highest in sectors relatively sheltered from international competition.

International comparisons suggest that the entry rate of new firms in Ireland is similar to those observed in Norway, Canada and the USA, although for different periods, but is perhaps 40% higher than the rate for the UK. Examination of new firm formation rates at *county* level reveals that high rates occur in a block of country south of Donegal and north of a line from Galway to Drogheda. By contrast, the counties of the South West, Mid West and South East have recorded low rates of new business formation. The analysis of new firm formation rates at county level, with number of new indigenous firms per annum per 1,000 employees in manufacturing (1973) as the dependent variable, suggests that there are two variables associated with this dependent: (1) the proportion of plants in the county employing fewer than twenty employees; and (2) the percentage of the county's population resident in towns of over 5,000 population. It appears that the number of new enterprises emerging is related to several features of the local area: the size distribution of firms; degree of urbanization; sectoral mix; and the rate of manufacturing employment change.

A relatively small proportion of new firms serve the sub-supply needs of the larger predominantly foreign companies: only 11.4% of the materials and components used by the largest New Industry sector, metals and engineering, were purchased in Ireland in 1976 (O'Farrell and O'Loughlin, 1981, p. 293) although the trend is upwards (O'Farrell, 1982). Much of the growth that has occurred in sub-supply industries has been in the lower skill areas, such as general welding, structural metal or packaging; indeed packaging represented over one-third of domestic purchases by the New Industry non-food sectors in 1976 (O'Farrell and O'Loughlin, 1980, p. 18). There are various supply side bottlenecks preventing the purchasing levels of foreign firms in Ireland rising to their potential of more than double the existing proportion (O'Farrell, 1982). These bottlenecks include lack of price competitiveness, an inability to achieve and maintain quality standards and unreliability in meeting delivery date deadlines (O'Farrell, 1982). Very few new Irish firms have been established in skill intensive sub-supply industries such as tool making, precision-casting, stainless steel valves or precision plastic moulds.

New metals and engineering firms constitute some 36% of all new businesses and most are engaged in general metal fabricating operations, metal bending and pressing, welding and repair shops, all of which typically serve local markets, and structural steel, where the economies also favour local suppliers. These businesses were founded as a response to increased demand created by plant construction, agricultural investment (farm gates and machinery) and general infrastructure expenditure (National Economic and Social Council, 1982). Employment generation has come almost exclusively from domestic demand; few firms have penetrated export markets. The principal opportunities for growth lie either in import substitution by providing components for MNEs or by firms currently serving only a regional or the national market exporting to the UK, the rest of the EEC or beyond. The major barriers to indigenous small firm growth are currently being investigated; they constitute a major policy problem for government and development agencies if the buoyant entry rates are to be translated into high manufacturing value added and employment. Although the rate of indigenous new firm formation in Ireland has been relatively high by international standards, and very high relative to the UK, there is evidence to suggest that most of the firms established are small concerns which are very unlikely to expand into even medium sized enterprises selling overseas.

Acknowledgements
The authors are extremely grateful to the Industrial Development Authority, Dublin, for providing the data upon which the analysis is based and for supporting the research and granting permission to publish the findings. Numerous IDA staff kindly provided information upon request; and Mr D. Flinter, Mr J. McMahon, and Mr M. Redmond contributed many useful suggestions upon an earlier draft. The normal disclaimer applies.

APPENDIX 1

Definitions

Establishment or plant
An identifiable unit of production engaged under a single legal entity in manufacturing activity at a distinct physical location. An establishment may be one of a number owned by a firm or enterprise but is classified separately if it has a discrete plant and work force at a specific location. Establishments may comprise one or more *technical units*: departments of a meat-packing plant which produce lard, cure bacon or canned meat are examples of technical units horizontally integrated within an establishment.

Enterprise or firm
A corporation, joint stock company, co-operative association, partnership, individual proprietorship or some other form of association. It owns and manages the property of the organization and receives and disposes of all its income; it may consist of more than one establishment.

Ownership status
The ownership variable has been classified into three categories: (1) multi-national branch plant; (2) Irish multiplant branch; and (3) indigenous (Irish) single plant. Joint ventures, of which there are only some twenty-five of the 5,000 plants, were classified under the majority shareholding group. Ownership in the case of surviving plants was categorized according to their status in 1981 and, for closures, their status in the year prior to closure. In the very small number of cases where a single plant firm expands by opening a branch, thereby moving into the multiplant category, or a multiplant firm disinvests and enters the single plant group, the plants are classified according to their 1981 status.

Openings
New establishments which were in existence in 1981 but not in 1973 and their employment 'gain', as a consequence of opening, is defined as their 1981 employment. The definition of the start-up date is, to some extent, arbitrary (see Mason, 1983). In this study it is defined as the year of entry to the IDA employment survey for which the qualifying criterion is a minimum full-time employment of three, including founder or partners.

Programme under which grant aided
Many grant-aided projects have received grant assistance under a number of separate programmes. In consultation with the IDA, it was decided to sort the programme classifications according to the following hierarchy. A *New Industry* grant takes precedence over all others: i.e. if a project has been in receipt of both a New Industry and a Small Industry grant, it is classified as New Industry for the purposes of the analysis. The remainder of the hierarchy, in order, is Small Industry, Re-Equipment, Shannon, Gaeltarra and Non Grant-Aided. Enterprise Development programme projects are classed as New Industry. A project is classified as grant-aided under one of the above programmes if at any time either before or after 1973 it has received a grant payment.

Relocations
Relocations are an unimportant phenomenon in Irish regional employment change. Only 139 plants, approximately 2% of the stock, relocated between 1973–81. The predominant movement was an inner city-suburban shift within the Dublin conurbation and fewer than fifty *jobs* actually migrated across a *regional* boundary. Hence, inter-regional relocations can be ignored in the analysis. It is extremely important, however, to identify and account for within-region relocations because the IDA files are assembled such that when a plant closes *prior* to relocating it is classified as a *closure*; and when it re-opens at a new site it is categorized as a *new plant opening* with a different numeric code. Hence, failure to identify and adjust for relocations *within* regions would seriously bias an analysis of new firm openings by inflating the number of new

firms and gross job increases arising from new openings. Relocations were identified by comparing alphabetical lists of possible relocations. A list of origins and destinations was then checked against a separate IDA listing of relocations and confirmed relocations were re-classified as permanent establishments for the purpose of deriving employment accounts.

APPENDIX 2

Definition of independent variables

X_1 = percentage of plants in industry i employing fewer than twenty persons in 1973

X_2 = rate of employment change of the 1973 stock of plants in industry i between 1973–81

X_3 = percentage of total employment in industry i located in plants employing over 200 persons in 1973

X_4 = median employment size of new indigenous single plant firms in industry i at end of first year

X_5 = median age of the 1973 stock of plants in industry i

X_6 = percentage of total employment in industry i controlled by multiplant firms in 1973

X_7 = percentage of plants in county j employing fewer than twenty persons in 1973

X_8 = rate of change in manufacturing employment in county j, 1973–81

X_9 = percentage of manufacturing employment in county j concentrated into plants employing over 200 persons in 1973

X_{10} = percentage of employment in county j in agriculture, 1971

X_{11} = percentage of employment in county j in commerce, retailing and wholesaling, 1971

X_{12} = percentage of population of county j living in towns of over 5,000 population, 1971

X_{13} = percentage of manufacturing employment in county j employed in indigenous single plant firms, 1973

X_{14} = median age of 1973 stock of plants in county j.

NOTES

1 Exclusion of firms which have never employed three or more people (including the owner-manager) will clearly produce a downward bias in new firm formation rates but will have only a marginal effect upon employment arising from new firm openings.

2 There are, however, other rural counties, notably Kerry, Tipperary S.R., Tipperary N. R., Carlow and Laois which, with respect to new firm formation, performed below the national average which suggests that it is not simply a rurality factor which explains the pattern.

3 For an extensive and thorough review of the contribution of economics to the understanding of new firm formation see Storey, 1982, pp. 47–74.
4 Due to a lack of adequate data, the profit variable cannot be tested in the empirical analysis.
5 The entry barriers literature is concerned with the prospects for self employment; scant consideration is given to the fact that a founder will usually move from an *existing* position of paid employment (or unemployment) and that, consequently, a subjective comparison of future earnings in paid and self employment will be made (Johnson and Darnell, 1976, p. 9)
6 The models fitted for survivors only data are not reported here but are available from the authors upon request. The parameters are very similar to those reported in Table 5.8.
7 The stability of the regression coefficients and standard errors suggests that multicollinearity is unproblematic.
8 Non-traded businesses include services localized within a region such as health care, public administration, retailing and house construction *and* manufactured goods in which the productivity improvements that can be achieved through increased scale are not great enough to offset the increased costs of distributing the product to a foreign country.
9 Models were also calibrated for survivors only and the results were very similar.

REFERENCES

Bain J. S. (1956) *Barriers to New Competition.* Harvard University Press, Harvard.
Churchill B. C. (1959) Rise in the business population, *Survey of Current Business Magazine*, May.
Collins L. (1972) *Industrial Migration in Ontario.* Statistics Canada, Ottowa.
Cooper A. C. (1971) Spin-offs and technical entrepreneurship, *I.E.E.E. Trans. Engin. Manag.* **EM–18**, 2–6.
Cross M. (1981) *New Firm Foundation and Regional Development.* Gower, Farnborough, Hants.
Dicken, P. and Lloyd P. E. (1978) Inner metropolitan industrial change, enterprise structures and policy issues: case studies of Manchester and Merseyside, *Reg. Studies* **12**, 181–97.
Fothergill S. and Gudgin G. (1979) The job generation process in Britain, Research Series 32, Centre for Environmental Studies, London.
Gudgin G. (1978) *Industrial Location Processes and Regional Employment Growth.* Saxon House, Farnborough, Hants.
Johnson P. and Cathcart D. G. (1979) New manufacturing firms and regional development: some evidence from the Northern Region, *Reg. Studies* **13**, 269–80.
Johnson P. and Darnell A. (1976) New firm formation in Great Britain, Working Paper 5, Department of Economics, University of Durham.
Mansfield E. (1962) Entry, Gibrat's Law, innovation and the growth of firms, *Am. Econ. Rev.* **52**, 1023–51.

Mason C. M. (1982) New manufacturing firms in South Hampshire: survey results, Discussion Paper 13, Department of Geography, University of Southampton.

Mason C. M. (1983) Some definitional difficulties in new firms research, *Area* **15**, 53–60.

National Economic and Social Council (1982) *A Review of Industrial Policy*, No. 64, NESC, Dublin.

O'Farrell P. N. (1982) Industrial linkages in the new industry sector: a behavioural analysis, *J. Irish Bus. Admin. Res.* **4**, 3–21.

O'Farrell P. N. (1984) Small manufacturing firms in Ireland: employment performance and implications, *Int. Small Bus. J.*, **2**, no. 2, 48–61.

O'Farrell P. N. and Crouchley R. (1983) Industrial closures in Ireland 1973–1981: analysis and implications, *Reg. Studies* **17**, 411–27.

O'Farrell P. N. and O'Loughlin B. (1980) *An Analysis of New Industry Linkages in Ireland.* Industrial Development Authority, Dublin.

O'Farrell P. N. and O'Loughlin B. (1981) New industry input linkages in Ireland: an econometric analysis, *Environ. Plann. A,* **13**, 285–308.

Robinson J. F. F. and Storey D. J. (1981) Employment change in manufacturing industry in Cleveland 1965–76, *Reg. Studies* **15**, 161–72.

Schumpeter J. A. (1961) *The Theory of Economic Development.* Oxford University Press, New York.

Storey D. (1982) *Entrepreneurship and the New Firm.* Croom Helm, London.

Wedervang F. (1965) *Development of a Population of Industrial Firms.* Scandinavian University Books, Oslo.

6

Innovation and regional growth in small high technology firms: evidence from Britain and the USA

R. P. OAKEY

INTRODUCTION

Small firm innovation and industrial growth

It is clear that innovation is the key to sustained prosperity and growth in the industrial firm (Mansfield, 1968; Freeman, 1974; Thwaites, 1978). Without improvements in product and process design in manufacturing, the competitive edge of the firm in national and international markets will decline over time (Feller, 1975; Ewers and Wettmann, 1980). This imperative applies throughout all sectors and scales of production, but will vary between industries and over time. There is evidence to suggest, however, that as individual industries mature, the scope for smaller producers is reduced due to formalization and subsequently increased cost of research and development, the standardization of products, and the general change in emphasis from product technology to process improvements (Vernon, 1966; Rees, 1979). With the general standardization of products, there is greater scope for the introduction of capital intensive machinery at scales and costs of production that exclude the small firm producer in, for example, motor vehicle production (Riley, 1973) or the electronics semiconductor industry (Rothwell and Zegveld, 1982).

Nonetheless, the broadly defined electronics and control instrumentation sectors are areas of high technology production where constantly evolving products and overall growth facilitates the continuing entry of many small firms to fill emerging new production niches (Oakey, 1981). Because the bulk of these evolving product niches are high technology, constant research and development is indicated by the need to keep internal product development at the leading edge of advancing technology. However, high technology small firms do not escape the typical problems of all small firms that emanate from birth and growth in conditions of finite financial resources. The conflict between the need for inevitably expensive R & D and a 'shortage' of investment capital is particularly intense in the small high technology firm, and may result in a defensive risk minimizing attitude, where an introverted approach to innovation is adopted in which firms move forward incrementally on the basis of internal profits (Boswell, 1973; Rothwell and Zegveld, 1982).

These resource problems notwithstanding, the small firm remains a particularly efficient vehicle for innovation in high technology electronics based industries where the informal juxtaposition of production with development ensures close interaction between concept and construction (Oakey, 1981; Rothwell and Zegveld, 1982). The impressive output from this type of organization is currently evident in the British computer industry where a significant proportion of recent innovations were in small firms (Townsend et al., 1981). Further evidence on the innovation potential of small firms emanates from the USA where large firms have acknowledged the advantages that a small scale of production brings to the innovation process through their investment in small firm joint ventures (Von Hippel, 1977; Roberts, 1977). While the potential for innovation in small firms may have been generally overstated in the media recently, innovation can be particularly cost effective in high technology small firms, provided there is an alleviation of the previously discussed financial constraints on R & D as a result of large firm or other financial assistance.

Small firm innovation, growth and regional development

The potential of small firms for national and regional employment growth has become a contentious issue since the release of the Birch Report findings (Birch, 1979). The service sector origin of many of the jobs discovered by Birch have been stressed by Fothergill and Gudgin, while they acknowledge that the *medium* employment potential of small manufacturing firms in regional economies can be substantial (Fothergill and Gudgin, 1979). Much of the current interest in small *high technology* firms lies in their potential for providing propulsive nodes of new high technology growth which act as embryonic vehicles for the industrial structural change of regions.

However, Storey (1982), in a study of employment within small firms in the Cleveland economy in the British Northern Development Region, indicated that, in the mainly low technology sectors of this area, employment growth of small firms over a ten year period was insubstantial. The Cleveland evidence suggests that the total number and sectoral type of new firms in a depressed region at any given moment will depend largely on the previous industrial structure of the local economy which is generally not conducive to the formation of new firms (Johnson and Cathcart, 1979). Nonetheless, if significant small-firm based employment growth is to be achieved in such depressed regions, it is most likely to emanate from the minority of small firms in fast growing high technology sectors. These firms might act as a basis for medium to long term significant *indigenous* employment growth through indigenous innovation (Oakey, 1983). It is certainly true that the originally small Texas Instruments Corporation played a central role in creating a high technology industrial agglomeration of human and material resources in less than twenty years in the Dallas–Fort Worth metropolitan area (Rees, 1977). While it has been known for many years that the growing industrial sectors of the British economy are concentrated in the South East of Britain (Cameron, 1979), much of the above discussion on the scope for industrial structural change in depressed regions must hinge on the ability of the adversely disproportionate share of growth sectors in such regions to create intra-regional structural change through indigenous small firm growth.

However, evidence from recent studies on regional variations in industrial innovation levels indicates that not only do depressed development regions in Britain possess a disproportionately low share of growth sectors, but when proportions are controlled and equal numbers of establishments are compared, plants in development regions are generally less innovative than their South Eastern counterparts. In particular it was discovered in two separate modules of research that small independent firms in the South East were almost twice as innovative as otherwise similar firms in the development regions (Oakey *et al.*, 1980, 1982). On the basis of these results, it was argued that the dependence of single plant independent firms on, by definition, a sole local industrial environment might explain these results. Clearly, the resource advantages of the local South East environment in, for example, public and industrial research and development (Buswell and Lewis, 1970), might explain the better innovation performance of the small firms in this region, while the poor local environment of the development regions might conversely inhibit innovation.

The results on which this paper is based stem from a research project undertaken to explore the regional effect of variable local input resources on innovation in small high technology firms. The main input resource foci of the study were based on previous research experience (Thwaites *et al.*, 1981), and concentrated on local finance, linkages, labour and technical information. This paper is concerned with an amalgam of data from the finance, labour and technical information sections of the study since they deal with the key inputs of human and financial research and development and innovation finance.

The survey

At a practical level, it was decided that the most rigorous means of testing the effect of local input resources on small firm innovation was to study the most diverse regional economic environments possible with hypothesized qualities ranging from the potentially least conducive to the most conducive for innovation. The Scottish development region was chosen as the potentially least conducive region and the San Francisco Bay Area of California (including Silicon Valley) was selected as the environment potentially most conducive for innovation. The South East of England, Britain's most innovative region, was also included as a region that would possibly be intermediate when compared with the other two environments, although it was more likely to be closest to the Bay Area in terms of quality and subsequent innovation performance. Since previous research has indicated that new product innovation was the most significant form of manufacturing innovation (Oakey, *et al.*, 1980, 1982), this measure was adopted in the current study to test the hypothetical innovation performance of survey regions. These general contentions on regional innovation levels are supported by Table 6.1 which indicates the incidence of product innovation in the three regions for the five year period prior to the survey. The hypothesized pattern of innovativeness between regions is generally confirmed both in terms of the least, intermediate and most innovative propositions with a 22% difference between the least innovative Scottish region and the most innovative Bay Area (Table 6.1)

The choice of a well known region of American high technology industry brought many advantages to the study of differing cultural and political attitudes. Moreover, there was some evidence that certain resources, notably information from universities (Deutermann, 1966; Cooper, 1970) and venture capital (Little, 1977) might be important in explaining the widely noted vigorous small firm growth in the Bay Area.

Various amalgamated directory sources were used in the British

Table 6.1 *Incidence of product innovations in the five year period prior to survey by region*

Product innovation	Scotland No.	%	South East No.	%	Bay Area No.	%
Innovation	34	63	47	78	51	85
No innovation	20	37	13	22	9	15
Total	54	100	60	100	60	100

Note: Chi square = 7·84, $p = 0·019$, $N = 174$.

regions to identify a universe, including Census of Production lists, the David Rayner Directory of Instruments and Electronics Manufacturers, and a Scottish Development Agency directory of electronics and instrument firms. The American data were derived from the California Manufacturers' Directory. The firms included were restricted to totally independent firms employing less than 200 workers, based on the high technology British instruments and electronics components Minimum List Headings (i.e. 354 and 364 respectively). However, although all the firms in the survey were operating in high technology sectors, firms were further categorized for the purposes of later analysis into three high, medium and low technology groupings based on the sophistication of their main product. The high technology grouping was mainly comprised of large laboratory instruments and control systems, while medium technology was typified by single function measuring and control devices and a small number of test equipment makers. The low technology category mainly consisted of printed circuit board, transformer and small electronic component producers.

The American firms were selected on the basis of the four digit Standard Industrial Classification codes that corresponded to the British MLHs. In practice, the process was relatively simple since the four digit American codes were virtually disaggregated to product level. In the South East and the Bay Area universes of appropriate firms were compiled, randomly sampled, and stratified by size with the use of replacement. In these two regions the sixty-firm target samples were achieved. However, the fifty-four firms forming the Scottish sample were the total population of co-operating firms in this region, and three less than the original universe of fifty-seven firms. If non-response is restricted to refusals, the overall response rate was a very satisfactory 79%. Identical questions were put to respondents in both Britain and the USA.

The following empirical sections of the paper deal in detail with
aspatial organizational and regional variations in the process of innovation in small high technology firms. This approach is divided into two
main sections; 'the nature of small firm research and development' and
'the role of innovation finance'.

THE NATURE OF SMALL FIRM RESEARCH AND DEVELOPMENT

Research and development cycles, product life cycles and risk

Apart from those small firms engaged in direct sub-contracting
arrangements, where items of production are specified to them by their
customers, small firms are predominantly engaged in the production of
an indigenously designed finished or semi-finished product. There is a
propensity in small high technology firms for products to enjoy
relatively short life cycles which are frequently as little as five years
duration. Thus it is clear that sooner rather than later existing products
must be improved or replaced to ensure continued sales and the
medium term survival and growth of the firm. Moreover, it is also
evident that in order for such periodic revisions in product specifi-
cations to be made, a firm must internally possess, or have access to, the
technical capacity to enact specification improvements. However, it is
inevitable that the research and development commitment, at whatever
level it is pitched, will be relatively costly.

A pervasive cause of investment problems associated with research
and development in high technology small firms is the cyclical nature of
both product sales and research and development expenditure. Most
high technology small firms are founded on a main product. The
growth and subsequent decline of sales over time from an initial or
subsequent new product implies that the profits will not be uniform.
This cyclical revenue will thus detract from the overall security of the
firm and its credibility with those providing risk capital. Moreover, to
exacerbate matters, Fig. 6.1 hypothesizes that maximum research and
development expenditure will consistently take place at times prior to a
new product launch when profits, for the purpose of the current
argument expressed *independently* of research and development costs,
are at a low ebb. In this conceptualization, the birth of the firm coincides
with low or non-existent profits, but high research and development
costs on an emerging product. This initial crisis at birth repeats itself at
intervals during the life of the firm, depending on the individual
product life cycles concerned. This simplified model assumes that all

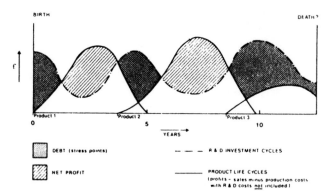

Fig. 6.1 Research and development investment and product life cycles in the small high technology firm

other innovation influencing variables, such as the performance of the economy, are held constant.

Fig. 6.1 indicates that points of financial stress occur when research and development expenditure exceeds profits. For a short period it may be necessary to obtain an overdraft or other loans with which to 'weather the storm'. In this sense, borrowing may be seen as a rather expensive means of smoothing the firm's profit curve, since future sales of the new product will, hopefully, finance the debts incurred during the stress periods. This problem of cyclical revenue is further compounded by the impact of taxation and its interaction with the product life cycle and ensuing 'lumpy' profits. Product life cycles will vary between firms, but in the conceptualization of Fig. 6.1 a product life cycle of approximately five years is adopted, a duration common in high technology small firms. For at least two of these years in mid–life cycle profits are high, thus attracting a high rate of taxation. However, during a period of loss or low profits there is no return of the tax that would have been deemed overpaid if averaged over the whole five year period. The net result of the combined negative factors of high interest on loans and taxation on profits is that the small firm has less money to spend on research and development directed at the development of future products.

Perhaps the biggest dilemma facing small firm owners relates to risk. Seen in a retrospective light, the success of the first and second products in Fig. 6.1 almost implies some form of symbiotic relationship between research and development effort and subsequent sales. However, lacking a crystal ball, the individual owner has no means of knowing the level of future sales when considering the appropriate level of research and development investment for any given new product development. He does know, however, the level of past expenditure on a product development at any given moment in the investment cycle. Since most small firms cannot afford the luxury of market research, much of the decision on how much to invest in a project and when to stop will stem from past experience and knowledge of current market conditions. While, by definition, all existing firms manage to ensure that research and development effort is broadly commensurate with subsequent sales potential, it is always possible that, as in the case of Product Life Cycle Three in Fig. 6.1, subsequent sales do not offset research and development expenditure, thus probably causing the demise of the whole firm. This phenomenon emphasizes that there is no justification for the argument that research and development investment is desirable at any price. This concept will be developed below in the context of investment capital.

Hence, a wide range of approaches and subsequent commitments to research and development are common in small firms. Clearly, firm size and the sophistication of the product technology will both influence the level of research and development commitment and must, at least in part, stem from the overall level of ambition inherent in the management approach (Senker, 1979). As a general rule of thumb, however, the quality of the technical information derived from research and development initiated in small firms is proportional to the total level of capital investment. For example, the most rewarding technical information on improved product design is likely to ensue from an internal research and development facility. Moreover, it is clear that the fruits from such efforts would be the confidential property of the firm concerned which may be important in competitive high technology markets (Oakey, 1981).

At a much lower level of commitment, a small firm might employ an external consultant on a contract by contract basis. This approach is clearly more flexible in that such expertise does not involve a full time research and development burden in employment terms. In extreme cases, a small firm might rely solely on such a consultant, but more frequently will use this service to support any given level of internal effort. The local availability of such a service would be a clear advantage to firms wishing to use such a facility, an issue of central concern to this paper. However, although flexible, problems with confidentiality may arise with external consultants and their work. Moreover, the quality of technical output from such a relationship will generally not yield the same quality of results as those produced by an internal research and development department (Oakey, 1979).

Internal research and development and innovation

Internal research and development facilitates the interactive linking of development effort to the specific production needs of the firm. In many small high technology firms the chief executive or owner is a qualified engineer. It is also likely that the *raison d'être* of the firm was the initial product idea of such an individual. Thus, for good or ill, there is rarely much conflict between specific technical research and development objectives and the wider aims of the firm, because the key decision maker is frequently the technical innovator (Senker, 1979, 1981). Virtually all high technology small firms with an internally developed product will make an effort to preserve their competitive edge through some form of evolutionary research and development. This assertion is borne out in Table 6.2, which conclusively indicates that only 26 (15%) of the 174 survey firms did not maintain some form of internal research and development facility. Moreover, the asserted link between internal research facilities and innovativeness, evident in earlier work (Thwaites *et al.*, 1981), is underlined by Table 6.3 which shows that the incidence of product innovation during the five year period prior to interview in survey firms with internal research and development was a high 93% level compared with a 7% proportion of innovators in the sub-sample without an internal research and development facility. From a regional viewpoint, it is significant that the largest number of firms with no research and development effort were located in Scotland.

However, the mere existence of research and development on site is an imprecise category which aggregates various degrees of full- and part-time commitments. Thus it is useful at this point in the discussion to consider full- and part-time research and development separately since these approaches imply different organizational structures within individual firms and varying levels of financial commitment.

Part-time research and development

Part-time research and development is particularly common in small firms. It is a compromise between the conflicting needs of innovation through research and development and the high cost of full-time research and development staff. Since in many small firms the owner is frequently the production engineer, accountant and salesman, great

Table 6.2 *The incidence of internal research and development by region*

Research and development effort	Scotland		South East		Bay Area	
	No.	%	No.	%	No.	%
Internal						
R & D	42	77·8	51	85·0	55	91·7
No R & D	12	22·2	9	15·0	5	8·3
Total	54	100·0	60	100·0	60	100·0

Note: Chi square = 4·31, *p* = 0·116, *N* = 174.

Table 6.3 *Presence of internal research and development by incidence of product innovation*

Research and development effort	Product innovation		No innovation	
	No.	%	No.	%
Internal				
R & D	123	93·2	25	59·5
No R & D	9	6·8	17	40·5
Total	132	100·0	42	100·0

Note: Chi square = 28·40, *p* = 0·0001, *N* = 174.

conformity of purpose is achieved by the owners' performance of development work in his spare time (Freeman, 1974). Indeed, in many instances, this practice precedes the establishment of the small firm. Work on the firm's initial product frequently takes place during a period of previous employment. Since full-time research and development workers are often a prohibitive expense, the part-time efforts of the owner, plus perhaps the technical director and production engineer may be an optimal level from a financial viewpoint. However, while part-time research and development would record a lower research and development cost-cycle in Fig. 6.1, the resultant output reflected in new product sales would also be lower.

In terms of the survey evidence, seventy-two (41%) of all firms claimed a part-time research and development effort. As might be expected, the majority of these firms were in the smaller size categories. Businesses employing less than fifty workers accounted for sixty-three of the seventy-two firm sub-sample. Moreover, only twelve (17%) of the seventy-two firms with part-time research and development were high technology firms. Thus in keeping with the generally modest character of this research approach, such fiᴏus tend to be mainly small and in lower technology areas of production. On a regional level, part-time research and development was least prevalent in Scotland where there was a greater tendency to either maintain full-time research and development (Table 6.4) or have none at all (Table 6.2). This heterogeneous Scottish pattern of results will be discussed in detail below. Over half the South Eastern sample performed research and development on a part-time basis, while the level in Bay Area firms fell between that of the two British regions.

Full-time research and development

Since full-time research and development, due to its relatively high cost, is rarely undertaken in a spend-thrift manner, its presence within firms implies a high level and quality of internally derived technical information. Indeed, the medium to long term benefits of research workers must be evident to ensure their survival (Oakey, 1981). The transition from an absence of research and development effort, through part-time status, to the establishment of full-time research and development clearly relates to firm size. With increasing firm size the owner/ key innovator role becomes impractical for the founder due to the demands of other administrative areas in the expanding business. This problem, plus the need for innovation, is ameliorated by the positive financial benefits of expansion, which, through profits, allows the employment of full-time research and development staff. This organizational change relieves pressure on the founder while still allowing his technical input to the full-time research if required.

Table 6.4 has indicated the high level of full-time research and development in Scotland within the sub-sample of plants with research and development. However, even after allowing for its lower proportion of plants with research and development, Scotland recorded the highest regional proportion of plants with full-time research and development expressed as a proportion of the *total* Scottish sample, with a 50% level compared with 33% and 48% for the South East and Bay Area regions respectively. Beyond the initial striking observation that the incidence of full-time research and development in Scottish survey firms is higher than in the Bay Area sample, and taking 1970 as a breakpoint, only thirteen of the twenty-nine American firms with full-time research and development were established after 1970 com-

Table 6.4 *Level of internal research and development commitment by region*

Research and development effort	Scotland No.	%	South East No.	%	Bay Area No.	%
Full-time	27	64·3	20	39·2	29	52·7
Part-time	15	35·7	31	60·8	26	47·3
Total	42	100·0	51	100·0	55	100·0

Note: Chi square = 5·93, $p = 0.051$, $N = 148$.

Table 6.5 *Firms with full-time research and development by age and region*

Age	Scotland No.	%	South East No.	%	Bay Area No.	%
Pre-1970	5	19	14	70	16	55
Post-1970	22	81	6	30	13	45
Total	27	100	20	100	29	100

Note: Chi square = 13·85, $p = 0.005$, $N = 76$.

pared with twenty-two of the twenty-seven firms in the Scottish Development Region (Table 6.5). Moreover, fourteen of the twenty-seven Scottish firms with full-time research and development were classified as high technology. The poor performance of the South East in terms of full-time research and development is partly vindicated by the region's good overall performance in full- and part-time research and development combined. Generally, however, the South East's performance is akin to that of the Bay Area since Table 6.5 indicates that the South Eastern relationship of firm age to full-time research and development facilities shows a majority of firms in the pre-1970 category, while in the Scottish instance the majority is clearly in the post-1970 grouping. Indeed, a significant minority of mainly high technology newer Scottish firms have produced a heterogeneous picture of innovation in Scotland, where a particularly innovative sub-sample of firms masks the performance of a notably backward group of firms to produce an indifferent overall regional innovative performance apparent in Table 6.1.

However, the mere occurrence of a full-time research and develop-

Table 6.6 *Percentage of total workforce in research and development by region*

R & D workers as a percentage of total employment	Scotland		South East		Bay Area	
	No.	%	No.	%	No.	%
1–9%	10	37·0	11	55·0	11	37·9
10–24%	13	48·1	7	35·0	13	44·8
25–49%	2	7·4	2	10·0	2	6·9
50% and over	2	7·4	0	0	3	10·3
Total	27	100·0	20	100·0	29	100·0

Note: N = 76.

ment facility might be a misleading measure of research and development effort. Various measures of the degree of the commitment to full-time research were available to the study, but financial measures such as research and development expenditure per worker in survey firms, which controlled for firm size, did not standardize for the higher wages paid in the USA for any given level of expertise. Since a major part of research and development costs in small firms is wages, a comparison on this basis would be invalid. A constant measure of research and development that again controls for firm size, but does not suffer the disadvantages of capital outlay, is research and development workers expressed as a proportion of the total workforce. Table 6.6, which utilizes this measure, confirms that not only does the Scottish region possess a significant minority of firms with full-time research and development, but that the levels of full-time commitment for any given percentage employment size band compares very favourably with the Bay Area.

External technical information and internal innovation

It is clear from the above analysis that the majority of high technology small survey firms maintain some form of internal research and development effort. Consequently, the general value of externally acquired technical information must be largely of a supportive nature in enhancing the overall internal research and development effort of the firm. Nonetheless, it is also true that an externally available source of technical information such as a university might give a significant boost to the internal research and development effort of small high technology firms. Indeed, it has been argued by writers in the USA that the concentration of high technology industry in both the Route 128 area near Boston and the Silicon Valley complex south of San Francisco owes much to the stimulation of local technically oriented universities (Deutermann, 1966; Gibson, 1970; Cooper, 1970).

However, much of the reasoning on the links between technical universities and local industrial growth depends on concomitance rather than direct causality. Certainly there is little hard evidence on locationally meaningful *information flows* between universities and local firms. It is often assumed that because local firms with technical needs and universities with relevant expertise are closely juxtapositioned, then interaction occurs. However, detailed evidence on the extent of links between high technology scientific instruments firms and universities in Britain indicated that contacts were infrequent and of a low technology supportive nature when they occurred (Oakey, 1979). Clearly, the current survey is an ideal opportunity to shed some light on possible internal differences in the technical links of firms with external research institutions.

Survey firm executives were asked if their firms maintained any external contact with a source of technical information of importance in developing products and processes in the plant. Surprisingly, Bay Area firms recorded the lowest level of external research and development links with only fourteen firms (23%) acknowledging a link compared with approximately half of the Scottish and South Eastern sub-samples (Table 6.7). There was evidence to indicate that local universities played an important part in producing the higher level of contact in Scotland. However, there appears to be little correlation between the most innovative region and extensive local and national links with external research institutions. Moreover, an attempt was made to ascertain the strength of the information flow in the minority of firms with a technical link to an outside body. Firm executives with links were asked if the breaking of the information flow, for whatever reason, would cause serious disruption to their internal innovation performance. A very small minority of only six Scottish, four South Eastern and three Bay Area firms responded affirmatively to this question.

These data give a clear overall impression that technical information links are not profuse, nor are those that do exist of great significance to innovation in small high technology survey firms. The impression created by the previous literature (Deutermann, 1966; Cooper, 1970) that American firms might maintain more abundant and technically important links with local and national technical universities is found to be false in terms of the small firms of the Bay Area sub-sample. Indeed, this result is particularly striking given that the majority of Bay Area

Table 6.7 *The incidence of important technical information contacts by region*

Technical	Scotland		South East		Bay Area	
contacts	No.	%	No.	%	No.	%
Important contacts	30	55·6	25	42·4	14	23·3
No important contacts	24	44·4	34	57·6	46	76·7
Total	54	100·0	59	100·0	60	100·0

Note: Chi square = 12·54, $p = 0·002$, $N = 173$.

firms were located in Silicon Valley, no more than ten miles from Stanford University. However, the importance of local American universities may be in their role of providers of 'spin off' entrepreneurs and skilled workers rather than in terms of interactive collaboration with existing firms (Zegveld and Prakke, 1978; Bullock, 1983).

THE ROLE OF INNOVATION FINANCE

Finance and the innovation process

Save in instances where research and development is performed on an informal basis by a key executive, the decision to conduct research and development in well organized firms must be accompanied by a decision on where the capital will be obtained with which to fund such efforts. Clearly, an element of risk is involved. Specifically, there are two separate but related risks associated with innovation finance in small high technology firms. First, there is a set of purely technical risks which accompany any manufacturing development. These are various and range from the danger of being beaten to the market place by competitors, to simply not achieving the level of quality that meets the specification requirement necessary to allow the product to function and sell in an envisaged production niche. Moreover, while success or failure is rarely total once products reach the market, there remains great scope for the prototype that does not meet minimum standards required for serious production. However, it is clear that failed products will not be widely publicized by the firm concerned for reasons of credibility.

The second set of risks confronting the small firm executive augment those on the technical viability of a new product. Even if the eventual technical viability of a new product development is assumed, the need to ensure the day-to-day survival of the business during the development of the product remains. In particular, the resources for the current products that produce internal funds for research and development must be provided. Clearly, in times of falling profits from ageing current products, cash flow problems may arise. In such circumstances, the firm may be faced with a stark decision. It can either reduce running costs through economies or it can borrow money to ease the firm through financial stress points in the innovation cycle caused by expensive research and development (Fig. 6.1). While individual sources of capital are considered in detail below, it is critical to stress that a new set of risks are introduced when a firm's management is forced to obtain external finance. Faced with the uncertainties associated with external borrowing, many small firm executives may consider the financial risks to the firm of continuing an expensive product development prohibitive and abort the project despite their confidence in the technical merits of the new innovation. For it is clear that in extreme instances, a costly research and development project that clearly will not bear fruit in the short term to aid cash flow may cause bankruptcy.

Since a central theme of this paper is the effect of local regional resources on the innovative performance of survey firms, it might be expected that variations in the availability of local external investment capital for expanding high technology firms would influence the extent and pace of innovation. The chosen study regions are well suited to this task since they are individual examples of diverse financial environments, ranging from the government assisted development region of Scotland, through the South East of England, with discretionary government assistance, to the Bay Area of California, perhaps the modern home of venture capitalism. The analysis of survey evidence in an investment finance context appropriately begins with an examination of regional variations in the availability of start-up capital and its effect on the birth of new high technology small firms.

Sources of start-up capital

The earlier arguments on product and research and development investment cycles (Fig. 6.1) imply that new firms are similar to existing small firms insofar as the traumas they experience at birth re-occur at intervals coincident with the introduction of subsequent new products, assuming that the new business in question prospers. In this sense, the initial product life cycle is distinctive only in that it may be the first of many such innovation processes during the life of the firm. However, it is clear that initial innovation in firms that produce a recognizable product from birth is particularly traumatic. The problem of investment finance is acute for new businesses because, by definition, they do not have access to the profits from previous sales with which to aid development. Moreover, the standing of the new firm with banks is exacerbated by the lack of a 'track record' on which to base a loan application. This problem has been noted by many distinguished investigations over the years (Macmillan, 1931; Bolton, 1971; Wilson, 1979).

Perhaps in view of the recent high interest rates in both Britain and the USA it is not surprising to note that 62% of all survey firms founded since 1970 relied on the personal savings of the founder for the initial injection of capital with which to begin the business. The sums involved may be very small and survey experience of conversations with founders suggest that in many cases the business is born with the aid of an increased overdraft of a few hundred pounds. Indeed, the business is often begun when the founder is employed at a large corporation; the full-time operation of the firm only taking place after a trial period in which sales and experience are accumulated. This incremental approach, although relatively slow, offers the major advantage of reducing the risks to the new firm founder when compared with faster loan based launches.

Regional evidence on the main source of start-up capital indicates the universal popularity of personal savings as a means of beginning a firm. The 62% average level of acknowledgement for this source is similar to levels observed in other studies of small firm 'start-up' finance (Cross, 1981; Storey, 1982). While a wide range of other sources are sparsely represented, the only other notable source of initial capital is local private venture capital (Table 6.8). It is clear that its status as the second most important source is caused by the Bay Area firms from which eight of the ten cases emanated. This is the first sign that the much vaunted venture capital market in the Bay Area is producing regionally different results within the total survey sample. Although numbers are relatively small, eight (30%) of the new Bay Area firms founded since 1970 were aided in their birth by venture capital, while this source was negligible in the British regions. This is a significant marginal regional effect which may aid the birth of small high technology firms in the Bay Area. However, it might be argued that the level of venture capital

Table 6.8 *Sources of start-up capital in firms founded since*
1970 by region

Start-up	Scotland		South East		Bay Area	
capital	No.	%	No.	%	No.	%
Personal savings	20	67	11	69	14	52
Previous assets	0	0	2	12	2	7
Bank loan	2	7	1	6	1	4
Second mortgage	1	3	0	0	1	4
Local venture capital*	1	3	1	6	8	30
Other	6	20	1	6	1	4
Total	30	100	16	100	27	100

* Within 30 miles of plant
Note: N = 73.

funding in the Bay Area is merely caused by the greater number of
innovative firms in this area reflecting greater scope for venture capital
funding. But this assertion is rendered unlikely by Table 6.1 which
indicated that the difference in product innovation levels between the
South East and the Bay Area are not great. The observed differences in
venture capital funding between the two nations is, therefore, more
likely to be a reflection of the variability of venture capital availability.

Subsequent employment growth

However, it should be stressed that small firms are, hopefully, transitory in nature. While there is clear evidence that many small firms
cease to operate within the first few years following their birth (Fothergill and Gudgin, 1982; Storey, 1981; Ganguly, 1982), it is also obvious
that, by definition, firms that grow cease to be small firms. Indeed, if
high technology firms in the sectors of this study grew rapidly between
1970 and the current survey date, they would not be potential study
firms due to the 200 employee ceiling placed on participants. Thus, the
following data on employment generated in new firms founded in the
five-year period prior to interview is a useful pointer to the job
generation powers of high technology small firms in the environmentally diverse survey regions. Moreover, the employment growth
of such new firms can be taken as a general surrogate for profitability
and success in the absence of regional comparable financial data, which
are difficult to present, both because of confidentiality problems surrounding profits, and difficulties in financial comparisons between
Britain and the USA.

Table 6.9 *Jobs created in firms less than five years old by region*

	Scotland	South East	Bay Area
No. firms	16	3	14
Total jobs	251	78	796
Average jobs per firm	16	24	57

Note: N = 33.

Table 6.9 indicates the number of survey firms founded in each region in the five-year period prior to interview with the total jobs created and the average of jobs per firm by region. It is surprising that the South East region recorded a mere three new firms during this period. Given that the sample from which these firms were drawn was randomly stratified by size and similar in other respects to its universe and the other regions, it is likely that these new firms are a fair reflection of small firm formation in the South East in the industries and period covered by the study. The other striking feature of Table 6.9 bears out observations made during interviews; in the Bay Area, firms appear generally to grow much faster than their British counterparts. If the Bay Area is compared with Scotland where the number of new firms are similar, the average size of firm is almost four times as large as the Scottish average figure.

However, the averages of Table 6.9 belie the true explanation for this sharp regional contrast in employment generation. In fact, the superior Bay Area performance is largely attributable to three new companies founded in the five-year period prior to survey that had grown to 125, 150 and 200 employees by the time of interview. The largest growth recorded in Britain was fifty-three employees in a South Eastern firm. This means that one Bay Area new firm generated jobs equivalent to 80% of the total jobs generated in sixteen Scottish new high technology small firms. Although cases are few in Table 6.9, this minority of fast-growing firms in the Bay Area did not exist in the other regions, and while the number of firms may be small, the jobs created are substantial. Moreover, there is every likelihood that other fast-growing firms in the Bay Area have grown beyond the 200 employees ceiling in the five-year study period. Indeed, certain discrepancies between the 1982 California Manufacturers' Directory employment figures for firms originally included in the survey, but subsequently excluded after contact because they had exceeded the 200 employees ceiling by mid-1982, anecdotally support this assertion.

It is generally true that the employment contributions of all small firms to regional economies may be of minor significance in the short term. However, Table 6.9 hints that in high technology firms within high technology agglomerations, the birth of a small number of fast-growing small firms may have a striking impact on employment. The importance of these minority results at the margin are leant added weight by the known performance of now famous high technology American firms that have grown from the type of new firm indicated in Table 6.9 into world famous electronics corporations such as Fairchild, Varian and Texas Instruments. All these firms are less than thirty years old – some much less; but their employment in Silicon Valley alone is measured in tens of thousands. With such impressive employment records, it is not necessary to generate hundreds of small firms employing an average of twenty employees each for the next fifty years, but merely two or three Texas Instrument type firms that subsequently become large.

Sources of investment finance

As noted in the case of new firm start-ups, it is no surprise to discover from Table 6.10 that there is an overwhelming tendency in all regions to move forward incrementally on the basis of internal profits when funding the main investment needs of the firm. However, Table 6.11 suggests that it is more frequently the highly innovative firm that seeks external funds, while Table 6.12 shows a tendency for these firms to be high or medium technology. Both these tables are statistically significant. Hence it appears that, due to the sophistication and innovativeness of the majority of external capital users, external sources of investment

Table 6.10 *The main source of investment capital by region*

Sources of investment capital	Scotland		South East		Bay Area	
	No.	%	No.	%	No.	%
Internal	35	64·8	49	81·7	45	75·0
External	19	35·2	11	18·3	15	25·0
Total	54	100·0	60	100·0	60	100·0

Note: Chi square = 4·25, $p = 0·120$, $N = 174$.

Table 6.11 *Incidence of product innovation by main source of investment capital*

Sources of investment capital	Product innovation No.	%	No innovation No.	%
Internal	93	70·5	36	85·7
External	39	29·5	6	14·3
Total	132	100·0	42	100·0

Note: Chi square = 3·86, p = 0·049, N = 174.

Table 6.12 *Technical complexity of main product by main source of investment capital*

Main source of investment capital	Technical complexity					
	High No.	%	Medium No.	%	Low No.	%
Internal	34	68·0	34	65·4	61	84·7
External	16	32·0	18	34·6	11	15·3
Total	50	100·0	52	100·0	72	100·0

Note: Chi square = 7·27, p = 0·026, N = 174.

Table 6.13 *Breakdown of main capital investment sources by region*

Capital sources	Scotland No.	%	South East No.	%	Bay Area No.	%
Profit	35	64·8	49	81·7	45	75·0
Local bank*	13	24·1	5	8·3	6	10·0
Local venture capital*	1	1·9	2	3·3	9	15·0
Other	5	9·3	4	6·7	0	0·0
Total	54	100·0	60	100·0	60	100·0

* Within 30 miles of plant
Note: N = 174.

capital may have an effect on the general level of innovation that is greater than its incidence in Table 6.10 would initially suggest. Moreover, this observation is of an enhanced relevance since a main aim of this paper is to investigate the effect of the local external environment on innovation.

There are two main sources of external capital available to survey firms. The first category which is available in both Britain and the USA includes government grants and loans. Clearly, while individual schemes differ between nations, and, indeed, between the two British planning regions, government funding is a broadly comparable category. The second category of external finance is private sector funding and, in terms of the current survey firms at least, divides into bank finance and private venture capital. These various detailed sources are indicated in Table 6.13 and initially considered briefly together in advance of individual sources.

It was apparent from Table 6.10 that Scottish and Bay Area firms indicated a marginally greater propensity to obtain investment capital externally than firms in the South East of England. However, the breakdown of these external sources in Table 6.13 indicates that they differ in that the main private external source in the Bay Area was banks *and* venture capital sources, while the dominant source of private capital in Scotland was the local banks alone. The residual 'other' category was mainly comprised of a mixture of various forms of British government assistance.

Government finance

It should be emphasized at the outset that few of the survey firms mentioned government incentives as a major source of investment capital for their firms. Thus the significance of government finance is more as a potential than actual source of investment finance. However, government incentives can provide a welcome boost to the overall resources of the firm, even when they are not the main capital source. This is particularly true if such assistance is in the form of grants or low interest loans. Hence there is, in theory, considerable scope for the use of government incentives in survey firms. In particular, in both Britain and the USA, there are incentives aimed specifically at small firms, while in the Scottish region, survey firms have access to additional regional development incentives available only in the British development regions. However, the impact of such schemes will be approached in terms of their *effective delivery* rather than discussing the

Table 6.14 *The incidence of government assistance by region*

	Scotland		South East		Bay Area	
	No.	%	No.	%	No.	%
Government aid	42	77·8	12	20·0	4	6·7
No Government aid	12	22·2	48	80·0	56	93·3
Total	54	100·0	60	100·0	60	100·0

Note: Chi square = 72·0, p = 0·0001, N = 174.

plethora of available schemes that theoretically exist to support small firms.

It is not surprising to discover from Table 6.14 that the Scottish sub-sample is by far the most extensive user of government incentives with a 78% level of incentive usage in the five-year period prior to interview. The figure for the South East was a very low 20% and an insignificant 7% for the Bay Area firms. The important role played in Scotland by regional development grants was confirmed by the discovery that thirty-five firms (65%) of the Scottish sub-sample had obtained regional development grants.

In many ways the South Eastern and Bay Area evidence is very similar as indicated by the low use of government incentives in Table 6.14. The reasons for not utilizing schemes in these regions were many, but the most frequently quoted reason for not having obtained government incentives in the past in both the South East and the Bay Area was that various forms of government red tape had either inhibited or caused the abandonment of attempts to obtain assistance. On the subject of red tape there were considerable similarities between the South Eastern firms and their Bay Area counterparts since most of the schemes available in these areas, unlike Scotland, were discretionary, and hence subject to a more extensive decision-making process. In the South East the Mapcon Scheme, designed to provide grants towards consultancy on possible micro-processor applications in the firm's products, had caused problems of access where, of the twenty-one firms attempting to obtain the incentive, only five had succeeded. In the Bay Area the pattern was much the same. The four survey firms in Table 6.14 obtaining incentives all made use of various Small Business Administration (SBA) loans. However, a further twenty-two firms in the Bay Area stated that they had sought, but failed to obtain, SBA loans. In instances where discretionary incentives were involved in both regions, the most common complaints concerned the bureaucracy involved in the application and the subsequently long time taken to make a decision on a loan.

Bank finance

On the surface there are substantial differences between the British and American banking systems. In particular, the British system is based on four main banks, while the United States, due to a legacy of restrictive legislation, has very few national banks and is instead served by many thousands of locally-based small banks. However, in practice, the international nature of banking practices in terms of both interest rates and loan conditions means that the effects of both British and American banks at plant level in the small firms of this study are very similar. For example, it is common knowledge that the politics of both the British and American governments in recent years have produced interest rates that are high by historical standards.

The plant level experiences of survey firms have been indicated in Table 6.13 which, apart from the Scottish instance with a 24% level of usage, confirms that the effects of banks as a main source of investment capital in survey firms is negligible in the South East (8%) and Bay Area firms (10%). The relatively high (24%) level of bank funding in Scotland may well reflect a genuine attempt by local Scottish banks in this region to play a more vigorous role in industrial financing than their nationally-based counterparts. This was certainly the view of a recent Monopolies Commission Report into the Royal Bank of Scotland (HM Government, 1982). However, the overall (14%) level of bank funding in survey firms is perhaps low given the generally high technology nature of the firms in the total sample. Conversations with individual executives in both Britain and America yielded a surprisingly similar range of comments on the value of bank funding. Indeed, many of the previous comments on the problems of government incentive delivery also apply to banks. It is often felt that the processors of loan applications are financially but not technically qualified personnel who do not see the sales potential of proposed innovations for which finance is required. And again, common criticism centres on the length of time taken to decide on a loan application. Also, there is a real fear of involvement with external agencies which, through loan agreements, might gain a say in company policy.

Venture capital

Venture capital and its effect on survey firms is worthy of inclusion here, not only to confirm its significant marginal effect on a minority of businesses, but because it is necessary to put in context many of the myths surrounding this form of investment finance. In terms of both start-up capital (Table 6.8) and as a main investment capital source (Table 6.13) it is clear that local venture capital is not a major source of investment capital in any of the survey regions. However, it is also abundantly clear in both tables that the pre-eminent origin of venture capital is the Bay Area of California. It may be recalled that 30% of the new firms founded in the Bay Area since 1970 were established with venture capital as a main source of start up capital, while 15% of the Bay Area total sample acknowledged that venture capital was a main source of investment finance during the five-year survey period. These marginal statistics gain added significance when compared with the virtual absence of any private venture capital sourcing in Britain.

Thus it is obvious that the impact of venture capital is limited in high technology small firms in the Bay Area and not pervasive to the extent claimed by many media commentators. This view is confirmed by other recent research in Silicon Valley (Bullock, 1983). However, it is also clear that venture capital does perform a useful function at the margin by making another source of investment finance available to entrepreneurs, particularly those with good prospective projects who are in need of a rapid injection of capital. It is certainly true that the nine Bay Area firms claiming venture capital as a main investment source were highly innovative. All the firms in this group had innovated in the five years prior to survey and eight of the nine presidents had an honours or doctoral degree. Moreover, this group also included the three fast growing firms mentioned above in the context of new firm start-ups, indicating the continuing role of venture capital in helping to explain the rapid growth of these firms since birth.

However, the most serious drawback to the involvement of venture capitalists in small high technology firms is the fear that internal control may be lost, at least in part, to the venture capitalists. This fear is more obviously manifested when equity is exchanged in return for investment capital, as is usual in such circumstances. Clearly, there is a danger that subsequent growth spurred on by the input of capital from previous equity sales may lead to further capital shortage which in turn leads to further equity transfer to the venture capitalist in return for more cash as the product life cycles increase in amplitude and consume greater resources than 'ploughed back' profits can provide as the firm grows. A point may be easily reached where the process is irreversible and the original small firm owner loses control of his own company, albeit with a healthy bank balance. Indeed, many venture capitalists are themselves entrepreneurs who have 'sold out' or been 'bought out' in the past, in this way accruing the experience and capital necessary to fund their venture capital organization.

CONCLUSION

It is clear from both the organizational and regional evidence of this paper that innovation in small high technology firms is best served by some level of internal research and development commitment. This internal imperative is confirmed by the general insignificance of external contacts with other sources of technical information in all regions which were used, when they occurred, in a mainly supportive role. However, the high cost of research and development was implicit in the wide spread of research commitments apparent in survey firms, ranging from the part-time effort of a single key innovator, to the full-time efforts of several engineers in a research and development department. While the medium term solvency of successful high technology firms is assured through their ability to pass on the frequently high cost of research and development to customers in the price of sophisticated high performance products, the short term financial standing of these firms can be uncertain due to costly research and development.

Evidence on sources of investment capital in survey firms tended to indicate that most small high technology firms moved forward incrementally on the basis of internally generated profits and relied on these earnings to fund whatever level of research and development they deemed necessary. However, there was evidence at the margin to indicate that innovative high technology firms were prevalent in the minority sub-group of firms utilizing external capital investment as a main source of investment capital. These external sources of capital were mainly comprised of institutional bank funding and private venture capital. It was notable that the Scottish banks were more successful in funding small firms when compared with their South Eastern or Bay Area counterparts. The marginal significance of venture capital in the Bay Area was given added weight by the virtual absence of any similar funding in the British regions.

An overall impression of regional variations in the data presented in this paper gives rise to two main observations. First, it is clear that Scotland has not readily accepted the role of a typical development region in decline. There is evidence to indicate that the heterogeneous sample of Scottish firms is mainly due to a mixture of old and new. In particular, the tendency for the Scottish sample to contain the highest regional incidence of both firms with no research and development, and businesses with a full-time research and development staff based on mainly newer high technology forms of production point to this internal diversity. The marginally higher incidence of bank funding, more frequent links with local universities, and the general ameliorative effects of the Scottish Development Agency in a supportive role through government incentives may all be partial causes of this encouraging Scottish performance. While it is too early to talk in terms of a true 'Silicon Glen' effect, this evidence on encouraging indigenous high technology small firm growth augurs well for future Scottish expansion in high technology electronics sectors. The second observation centres on venture capital. Apart from the overt spatial difference in venture capital availability between Britain and the United States, the above observations on venture capital contain valuable policy implications. Hence, venture capital funding forms a basis for the following discussion of policy issues raised by this paper.

There is evidence that it is mainly fast growing firms that vigorously seek external investment capital. Moreover, limited data from the Bay Area indicates that venture capital organizations are particularly attracted to this type of firm, as might be expected. However, from a British policy viewpoint, it is these same types of firms at which government finances might be particularly directed to fill the vacuum caused by the absence of venture capital. This vacuum is particularly acute due to the observed concomitant generally poor level of institutional private sector venture capital funding on the part of British banks in the survey. If it is accepted that there is scope for government agencies in a venture capital role, there are aspects of venture capitalist behaviour that might act as a model for public sector extenders of assistance in an attempt to fill this venture capital 'gap' in the British context.

It has been noted that government incentive delivery, in the more prosperous and innovative South East and Bay Area regions, is largely ineffective. Much comment was made by executives during interviews on the length of time taken to obtain finance, or at least a decision, and on the technical incompetence of government and banking bureaucrats who were predominantly versed in financial matters only. Significantly, venture capitalists are generally successful because they avoid these problems. They are mainly ex-businessmen themselves with both business and technical acumen. Moreover, if they decide to back a project, the cash is provided quickly.

Clearly, many of these strengths could be incorporated in a new approach to government funded innovation finance directed at fast growing high technology small firms. In particular, the need for technically qualified decision makers on loan application is essential, together with a collapsed time scale for the decision process. On a more general level, it is clear from the evidence of this paper that innovation in high technology small firms is at times both lengthy and expensive. Thus, with the confidence stemming from technically based judgements, loan agencies should be prepared to back firms over an extended time span (perhaps five years). This backing could take the form of an exchange of capital for an equity stake. The cost might be high, but so too can be the medium term return on investment. As in the venture capitalist instance, risk to the loan agency could be reduced by the development of a portfolio of firms (Bullock, 1983).

Within any region, regardless of its relative vitality, there will be a number of emerging high technology small firms. If government bodies working at a local regional level were empowered to take an equity stake in emerging small firms with potential in return for investment finance, such indigenous small firms could be encouraged to grow, particularly in depressed areas, where industrial structural change is most needed. This regional approach is in some ways a similar role currently performed by the National British Technology Group (BTG). Equity might be returned to the company at a later date if required, but this approach would stimulate indigenous growth and, importantly, protect embryonic business in development regions with good potential products from the product asset stripping of large competitors who might relocate the acquired product to another region after purchase of the firm (Smith, 1982). It must be emphasized, however, that this is not a 'lame duck' approach and would deliberately focus on high technology small firms with potential.

Conversely, it might be argued that the evidence of this paper has indicated that any form of government involvement would be resisted by small firm entrepreneurs. However, the validity of this view is diminished by the success achieved by both the British Technology Group and the Scottish Development Agency in their funding of small firms. This success demonstrates that government assistance to small firms can be effective when it is delivered by credible organizations with proven business and technical expertise. Entrepreneurs will show greater enthusiasm and respect for the agents of government assistance if they can prove their understanding of small firm problems and potentials and do not suffocate the aid process with irrelevant forms and unnecessary delays. The respondents to this survey in both Britain and the USA were rarely adverse to government aid *per se*, but merely abhorred the current over-bureaucratic system of delivery. Under the proposed venture capital agency approach, the exchange of equity for finance, together with an acceptance of a nominated agency director on the company board, will be bearable if the firm owners feel that they are augmenting the overall organizational, technical and financial strength of the firm.

In the fast growing, high technology sectors of this study one new firm in a depressed area with the growth potential evidenced in the past by Fairchild or Texas Instruments could change the image and employment structure of a region within ten years, if correctly nurtured. It would be in the national and regional interest for government agencies given the remit of promoting industrial growth to recognize and encourage the indigenous growth of such firms. The conduciveness of local regional resource environments for small high technology firm growth is clearly a significant partial cause of overall regional growth, and investment finance is a critical factor in such a growth process. Currently, there is a discernible regional difference between the financial environments of Britain and the USA with regard to the funding of firms with rapid growth potential. However, since most governmental assistance is financial, there is no reason why the local resource environment, particularly in depressed regions, could not be improved by the resort to new small firm investment policies directly aimed at promoting and supporting research and development and subsequent innovation in small high technology firms, as part of a wider strategy for industrial structural change in depressed regions.

REFERENCES

Birch D. L. (1979) The job generation process, Working Paper, MIT Program on Neighborhood and Regional Change, Cambridge, Mass.

Bolton, J. E. (1971) *Small Firms: Report of the Commission of Enquiry on Small Firms*, Cmnd. 4811. HMSO, London.

Boswell, J. C. (1973) *The Rise and Decline of the Small Firm*. George Allen and Unwin, Hemel Hempstead, Herts.

Bullock M. (1983) *Academic Enterprise, Industrial Innovation and the Development of High Technology Financing in the United States*. Brand Brothers, London.

Buswell R. J. and Lewis E. W. (1970) The geographic distribution of industrial research activity, *Reg. Studies* **4**, 297–306.

Cameron, G. C. (1979) The national industrial strategy and regional policy, in Maclennan D. and Parr J. B. (Eds.) *Regional Policy: Past Experiences and New Directions*, pp. 297–322. Martin Robertson, Oxford.

Cooper A. C. (1970) The Palo Alto experience, *Ind. Res.* **12**, 53–5.

Cross M. (1981) *New Firm Formation and Regional Development*. Gower, Farnborough, Hants.

Deutermann E. P. (1966) Seeding science based industry, *New England Bus. Rev.* December, pp. 7–15.

Ewers H. J. and Wettmann R. W. (1980) Innovation oriented regional policy, *Reg. Studies* **14**, 161–79.

Feller I. (1975) Invention, diffusion and industrial location, in Collins I. and Walker D. F. (Eds.) *Locational Dynamics Manufacturing Activity*, Wiley, Chichester.

Fothergill S. and Gudgin, G. (1979) The job generation process in Britain, Centre for Environmental Studies Research Series 32, London.

Fothergill S. and Gudgin G. (1982) *Unequal Growth*. Heinemann, London.

Freeman C. (1974) *The Economics of Industrial Innovation*. Penguin, Harmondsworth.

Ganguly A. (1982) Significant surplus of births over deaths, *British Business* **23**, 512–13.

Gibson J. L. (1970) An analysis of the location of instrument manufacture in the United States, *AAAG* **60**, 352–67.

HM Government (1982) Royal Bank of Scotland, Hong Kong and Shanghai Banking Corporation, Standard Chartered Bank – a report on the proposed merger, Cmnd. 8472, HMSO, London.

Johnson P. S. and Cathcart D. G. (1979) New manufacturing firms and regional development: some evidence from the Northern Region, *Reg. Studies* **13**, 269–80.

Little A. D. (1977) New technology based firms in the United Kingdom and Federal Republic of Germany, report prepared for the Anglo-German Foundation for the study of industrial society.

Macmillan Committee (1931) *Report of the Committee on Finance and Industry*, Cmnd. 3897, London, HMSO

Mansfield E. (1968) *The Economics of Technical Change*. Longman, London.

Oakey R. P. (1979) The effects of technical contacts with local research establishments on the location of the British instruments industry, *Area* **11**, 146–50.

Oakey R. P. (1981) *High Technology Industry and Industrial Location*. Gower, Aldershot, Hants.

Oakey R. P. (1983) New technology, government policy and regional manufacturing employment, *Area* **15**, 61–5.

Oakey R. P., Thwaites A. T. and Nash P. A. (1980) The regional distribution of innovative manufacturing establishments in Britain, *Reg. Studies* **14**, 235–53.

Oakey R. P., Thwaites A. T. and Nash P. A. (1982), Technological change and regional development: some evidence of regional variations in product and process innovation, *Environ. Plann. A* **14**, 1073–86.

Rees J. (1977) Location decisions, linkages and industrial growth, unpublished PhD thesis, University of London.

Rees J. (1979) Technological change and regional shifts in American manufacturing, *Prof. Geog.* **31**, 45–54.

Riley R. C. (1973) *Industrial Geography.* Chatto and Windus, London.

Roberts E. G. (1977) Generating effective corporate innovations, *Technol. Rev.* **80**, 27–33.

Rothwell R. and Zegveld W. (1982) *Innovation and the Small and Medium Sized Firm.* Frances Pinter, London.

Senker P. (1979) Skilled manpower in small engineering firms: a study of UK precision press tool manufacturers, Science Policy Research Unit, University of Sussex.

Senker P. (1981) Technical change, employment and international competition, *Futures*, June, pp. 159–70.

Smith I. (1982) Some implications of inward investment through takeover activity, *Northern Economic Review* **2**, 1–5.

Storey D. J. (1981) New firm formation, employment change and the small firm; the case of Cleveland County, *Urban Studies* **18**, 335–45.

Storey D. J. (1982) *Entrepreneurship and the New Firm.* Croom Helm, London.

Thwaites A. T. (1978) Technological change, mobile plants and regional development, *Reg. Studies* **12**, 445–61.

Thwaites A. T., Oakey R. P. and Nash P. A. (1981) Technological change and regional development in Britain, final research report, Centre for Urban and Regional Development Studies, University of Newcastle upon Tyne.

Townsend J., Henwood F., Thomas G., Pavitt K. and Wyatt S. (1981) Innovations in Britain since 1945, Occasional Paper Series No. 16, Science Policy Research Unit, Sussex.

Vernon R. (1966) International investment and international trade in the product cycle, *Quart. J. Econ.* **80**, 190–207.

Von Hippel E. (1977) Successful and failed co-operative ventures; an empirical analysis, *Ind. Market. Mgmt.* **6**, 163–74.

Wilson Committee (1979) *The Financing of Small Firms*, Interim Report of the Committee to review the functioning of the financial institutions, Cmnd. 7503, HMSO, London.

Zegveld W. and Prakke F. (1978) Government policies and factors influencing the innovative capability of small and medium enterprises, paper presented for the Committee for Scientific and Technological Policy, OECD, Paris.

7

Regional variations in capital structure of new small businesses: the Wisconsin case[1]

RON E. SHAFFER and GLEN C. PULVER

INTRODUCTION

Despite the increasing evidence of the importance of new business formation, and the importance of small businesses to the economies of various areas (Armington and Odle, 1982; Birch, 1979; White, 1980), there is minimal amount of information available on how people assemble the capital necessary to initiate the new business (Andrews and Eiseman, 1981; Daniels and Lirtzman, 1980; Shapero, 1983). There is a general lack of information regarding the ability of new small businesses to obtain financing, and the extent of differences in financing is influenced by where the business is located (Andrews and Eiseman, 1981; SBA 1983; Dunkelberg and Scott, 1983; Shapero, 1983). If there are substantial differences, the financial institutions concerned with small businesses need to be aware of these differences, so the flow of capital is not unnecessarily impeded.

Reports are beginning to appear that purport to show that there is or is not a capital shortage for small businesses (CNEA, 1981; Daniels and Lirtzman, 1980; SBA, 1983). The conclusions seem to be sensitive to how a capital gap is defined and national economic conditions.

One of the more comprehensive surveys of capital needs of small businesses was completed by the Council for Northeast Economic Action in 1980 (CNEA, 1981). They found that a capital gap existed for 1.1–1.7% of the financially sound firms, defined as debt/equity ratio of one or less. This amounted to 46,000 to 71,000 firms in the US. They found the degree of unmet capital needs increased with rural locations, desire for unsecured intermediate and long-term debt, and declines in firm size and increased leveraging. Daniels and Lirtzman (1980) in their study argue that the cost of capital is relatively less important than the

166

availability of capital to small firms. They then contend that small firms do not have the same degree of access to the capital market as do larger firms.

This study does not intend to determine the presence or absence of a capital gap, but does intend to address a dimension missing in most previous work. Most prior work examines the use of capital by firms established and operating for a number of years. This study will focus on the new small firm during the first few years of its existence. If small firms are important to employment generation and innovation and development of new products and markets then how they are capitalized and difficulties they maý experience in financing are important policy issues (SBA, 1983).

The purpose of this study is to examine the functioning of the capital market for new small firms in a thinly populated region compared to a heavily populated region and in urban or rural locations in Wisconsin. The functioning of the capital market will be examined through similarities and differences in the financial structure, credit denial experiences, and perceptions of capital availability among firms in the different regions.

The thinly populated region selected for this study is the northern part of the State of Wisconsin in the United States. It is typical of the heavily forested portion of the Great Lakes Region of the US. People are generally reliant on small farms, small businesses, forestry, and tourism for their income and employment. For the purposes of this study it is identified as the Upper Great Lakes Region (UGLR). This is contrasted with the more heavily industrialized, densely settled portion of southern Wisconsin (non-UGLR). These two areas are further subdivided into rural and urban locations.

A comparison is also made among urban and rural settings in the same state. Urban refers to a business located within thirty miles of a city of 30,000 or more. This definition is used to capture differences in the functioning of small business capital markets. A city of 30,000 should have relatively well developed financial institutions arising from factors such as size of place, competition, and personal contact among members of the financial community. Thirty miles is hypothesized as a distance people would drive to or be knowledgeable about financial institutions in a city of 30,000 or more.

On the following pages the general hypotheses examined will be explained. Then the types of firms surveyed along with their financial structure among regions will be reviewed. The debt and equity capital sources for the firms interviewed follows that section. The extent of and firms' response to credit denial are summarized before examining a summary indicator of capital stress.

THE HYPOTHESES

Modern capital market theory and business finance theory suggest a number of hypotheses regarding the financing of new small businesses (Daniels and Lirtzman, 1980; Mikesell and Davidson, 1983). These can be tested by the data collected for this study. First, the competitive capital market will make appropriate judgments in the allocation of capital over space and uses. This means that the sum of individual decisions in the capital market yields the optimal use of capital among places such as rural or urban and more or less developed areas. Furthermore, capital is allocated among uses to maximize the output of the economy, e.g. speculative oil drilling versus purchasing new machinery and equipment. While this study cannot make final judgments on either of these hypotheses, it can provide some insight to the allocation of capital among different economic regions. A second hypothesis is that the decision making capacity of financial institutions is not an impediment to the appropriate flow of capital in the economy. If there is an information problem in the capital market, as more information is acquired the access to capital should improve, other things given. This means that as 'new' firms compile a record of operation they will have improved access to more and different forms of capital. A final general hypothesis relevant for this study is that the size of the firm will have no effect on access to the capital market. Size can also take the form of rapid versus less rapid growth.

THE SURVEY

The data required for this analysis was collected through personal interviews in 1979–80 with the owners of new small businesses. For purposes of this study small business can be defined as (1) independently owned and operated; (2) not dominant in its field; (3) able to be overseen and managed with a great deal of knowledge by one person; and (4) generally employing less than 150 people. A firm was defined as new if it did not exist, under current ownership, prior to January 1976.

Nine Standard Industrial Classification (SIC) groups were chosen to represent a broad spectrum of business types ranging from retail to manufacturing (see Table 7.1).

Approximately two-thirds (67.2%) of the firms surveyed were

Table 7.1 *Standard Industrial Classification of the 134 firms interviewed*

SIC	Description of SIC	Number of firms in sample	Number of firms in sample by region					
			UGLR		Non-UGLR			
			Total	Rural only	Total	Rural only	Rural	Urban
15	Construction – general contractors and operative builders	15	10	6	5	1	7	8
354	Metal working machinery	16	3	1	13	1	2	14
355	Special industry machinery							
356	General industry machinery							
3599	Misc. machinery except electrical							
421	Trucking, local and long distance	16	13	6	3	–	6	10
508	Wholesale trade, durable goods – machinery, equipment and supplies	17	12	5	5	3	8	9
514	Wholesale trade, nondurable goods – groceries, and related products	8	5	3	3	–	3	5
541	Retail grocery stores	16	13	6	3	2	8	8
571	Retail furniture and home furnishing stores	13	12	8	1	1	9	4
58	Eating and drinking places	20	16	11	4	–	11	9
592	Retail liquor stores	13	7	5	6	1	6	7
	Total	134	91	51	43	9	60	74

located in the Upper Great Lakes Region (UGLR). Subdividing UGLR and non-UGLR further, 56% and 21% of the firms were in rural locations respectively. Using the rural/urban dichotomy, statewide 44.8% of the firms interviewed were in rural locations.

The names and addresses, by SIC code, for 432 businesses with new employer identification numbers during 1977 and less than 150 employees were obtained from Unemployment Compensation records of the Wisconsin Department of Industry, Labor and Human Relations. Of the 432 firms initially included in the sample, 121 (28%) were disqualified because they had a start-up date prior to January 1976, for 97 (23%) of the firms the owner could not be located and the firm was assumed to be nonoperational, 16 firms yielded incomplete interviews and 134 (31%) of the original sample yielded useable interviews.

Table 7.1 indicates there is substantial variation in the types of firms interviewed in the different areas of the state. This reflects the economic structure of those areas. While some of the geographic differences in the following discussion may arise from types of firms in the area, a preliminary analysis of the capital structure of the firms (Table 7.2) suggests that is only a partial explanation. Furthermore, economic regions are not spatially separated replications of some 'standard' economic structure.

DESCRIPTION OF FIRMS

The responses in the interview dealt with financial activities beginning with each firm's start-up date through December 31, 1978. The start-up date of the businesses interviewed ranged from January 1, 1976, through December 31, 1977. On December 31, 1978, the average age of the firms interviewed was 23.8 months in UGLR, 25.0 months in non-UGLR, 25.3 months in rural areas and 23.1 months in urban areas.

Sole proprietorships accounted for at least half of the firms (54%) and corporations were the next most popular form of organization (38%).

Table 7.2 presents information on the capital structure of the firms interviewed when categorized by the industry type. Part of the diversity among the businesses interviewed appears to be related to the type of business. Start-up employment ranged from 3.39 workers in retail liquor stores to 9.64 workers in wholesale establishments. Wholesale establishments experienced the greatest employment growth rate (92.5%) and eating and drinking places the lowest (14.6%). Construction firms started with the smallest average total assets, but grew the fastest between start-up and December 1978. Wholesale firms had the

Table 7.2 *Financial and employment structure of surveyed firms: by sector*

	Total employment		Total assets		Total debt		Net worth		Debt to net worth[a]	
	Start-up	Dec. 78	Start-up ($)	Dec.78 ($)	Start-up ($)	Dec. 78 ($)	Start-up ($)	Dec. 78 ($)	Start-up	Dec. 78
Construction (n = 15)	4.13	6.20	19,173	47,667	19,567	40,750	-394	6,917	.83	4.43
Manufacturing (n = 16)	3.69	6.94	75,249	137,021	43,020	84,748	32,229	52,273	.38	2.38
Trucking (n = 16)	3.44	4.81	57,138	130,644	36,750	87,738	20,388	42,906	3.85	3.73
Wholesale (n = 25)	9.64	18.56	160,276	384,913	126,350	296,028	33,926	88,886	4.04	2.33
Retail (n = 29)	8.83	11.31	97,975	158,310	63,621	88,323	34,354	61,080	3.50	1.70
Eating and drinking (n = 20)	8.90	10.20	66,840	87,200	48,115	41,900	18,725	45,300	2.41	1.88
Liquor (n = 13)	3.39	4.23	90,192	122,915	57,077	77,212	33,115	45,704	2.82	2.32

[a] The debt to net worth ratio reported is the average of the individual firms, not the ratio of average debt to average net worth.

largest start-up total assets and also had the second fastest rate of growth. Construction, trucking, and wholesale firms doubled their total assets between start-up and December 1978. Firms with high rates of growth in total assets also had similar rates of growth in total debt. The major exception was the 12.9% decline in total debt of eating and drinking establishments. Construction, trucking, and wholesale firms more than doubled their debt between start-up and December 1978. Construction firms had the smallest average debt both at start-up and December 1978. Wholesale firms had the largest average debt both at start-up and December 1978. Construction firms averaged a negative net worth at start-up, but experienced a phenomenal increase over the first few years of existence of the firm. Trucking, wholesale, and eating and drinking establishments more than doubled their net worth during the first 1–3 years of their operation. A 38.0% increase for retail liquor stores was the smallest increase in net worth among the different types of industries. The debt to net worth ratio for manufacturing firms increased from .38 to 2.38 or 526%. Construction firms had a 434% increase in their debt to net worth ratio. While these two industries were increasing their use of debt, the others reduced their use of debt relative to net worth. The retail grocers and furniture dealers halved their use of debt relative to net worth.

Table 7.2 indicates that there are differences in capital structure arising from the type of business. This may create some bias in examining regional variation (particularly UGLR versus non–UGLR, but not rural versus urban), but the samples reflect the type of economic activity in each area. The size of the total sample prevents fuller exploration of both economic sector and geographic region simultaneously.

Firms in the UGLR and urban locations averaged more total assets than firms in the other areas, and also had a larger increase in total assets between start-up and December 1978 (see Table 7.3). Firms located in the UGLR and in urban areas had a greater average debt at start-up and December 1978. The rate of growth of debt was higher for firms in the UGLR and rural firms suggesting these firms were able to acquire debt once they started. The firms in the sample exhibited substantial increases in net worth. Between start-up and December 1978 the average net worth of the firms interviewed doubled. Firms in the UGLR and in urban locations displayed a higher rate of increase in net worth than did the remaining firms. The net worth for firms in rural locations averaged at least 25% more than for firms in urban locations even though total assets averaged less, both at start-up and December

Table 7.3 *Balance sheet composition by area*

	Start-up amount ($)	Dec. 1978 amount ($)	% change[a]
UGLR (n = 91)			
Total assets	96,271	185,121	92.29
Total debt	69,045	126,891	83.78
Net worth	27,226	58,230	113.87
Non-UGLR (n = 43)[b]			
Total assets	69,828	132,120	89.20
Total debt	46,391	84,250	81.61
Net worth	23,436	46,447	98.19
Rural (n = 60)[c]			
Total assets	85,892	160,747	85.85
Total debt	51,647	97,962	89.76
Net worth	34,244[d]	61,624	79.94
Urban (n = 74)			
Total assets	89,321	174,473	95.33
Total debt	69,987	125,636	79.51
Net worth	19,344[d]	48,837	152.60
Total (n = 134)			
Total assets	87,785	168,384	91.82
Total debt	61,775	113,245	83.32
Net worth	26,010	54,138	108.15

[a] There was a significant change (at 10% level) in the start-up and Dec. 1978 average for all balance sheet elements for all regions.
[b] 1978 average based on n = 42.
[c] 1978 average based on n = 59.
[d] 10% significant difference in the means between regions.

Table 7.4 *Financial structure of UGLR and non-UGLR firms classified by rural/urban location*

	Total assets ($)		Total debt ($)		Net worth ($)		Debt to net worth[a]	
	Start-up	Dec. 78	Start-up	Dec. 78	Start-up	Dec. 78	Start-up	Dec. 78
UGLR								
Rural	85,046	154,207	50,850	94,180	34,196	60,026	3.24	1.86
Urban	110,584	224,537	92,244	168,597	18,340	55,940	2.21	3.49
Non-UGLR								
Rural	90,684	202,438	56,167	119,394	34,517	49,439	2.08	1.72
Urban	64,307	115,574	43,804	75,095	20,503	40,480	2.79	2.64

[a] The debt to net worth ratio reported is the average of the individual firms, not the ratio of average debt to average net worth.

1978. The difference in average net worth between UGLR and non-UGLR firms was much less pronounced.

Table 7.4 further refines the data for UGLR and non-UGLR into rural and urban. The rural firms in the UGLR had fewer total assets and total debt at start-up and December 1978 than did the urban firms in the UGLR. However, the net worth of UGLR rural firms was greater than for UGLR urban firms. Rural UGLR firms actually reduced their use of debt relative to net worth between start-up and December 1978, while UGLR urban firms increased their use of debt to net worth. Outside the UGLR areas the rural firms were larger in start-up and December 1978 total assets, total debt, net worth, and 1978 gross sales. While both rural and urban firms in the non-UGLR area reduced their use of debt relative to net worth, rural firms reduced their use more and had started with a smaller debt to net worth ratio.

Total employment for the UGLR and urban firms averaged more both at start-up and December 1978 than did the remaining firms (see Table 7.5).[2] The change in total employment for each region was statistically significant at 10%. The data in Table 7.5 indicates that the employment increase in these firms was through the generation of full-time jobs rather than part-time jobs or increases in owner and unpaid family labor. Thus, these firms were creating the full-time jobs a community is likely to seek.

The information on debt and net worth in Table 7.3 can be combined into a debt to net worth ratio. The debt to net worth ratio provides a measure of ability to leverage owners' funds with borrowed or debt capital. A higher ratio indicates an ability to acquire debt funds or an attitude supporting the use of debt capital.

At start-up, firms in the UGLR averaged at least one additional dollar of debt per dollar of net worth (owner's equity) than did firms located outside the UGLR (Table 7.6).[3] The debt to net worth ratio of rural firms averaged only slightly larger than that of urban firms. This, coupled with the greater net worth for UGLR and rural firms, suggests debt capital is at least as available as in other locales.

While the trend in the debt to net worth ratio was downward, it varied widely among geographic locations and even increased in urban locations (Table 7.7).

At start-up the average debt to net worth ratio for firms in each region was more than three. A larger percentage of firms in the UGLR had leveraged their equity more (> 3) compared to firms in the non-UGLR at start-up (see Table 7.6). A larger percentage of firms in the non-UGLR had a start-up debt to net worth ratio of zero, sugges-

Table 7.5 *Average employment by region*

| | Owner and unpaid family | | | Paid workers | | | | | |
| | | | | Part-time | | | Full-time | | |
	Start-up	Dec. 78	% change[a]	Start-up	Dec. 78	% change[a]	Start-up	Dec. 78	% change[a]
UGLR	1.82	1.77	-3.02	2.62	3.80	45.39	2.15	5.43	148.9
Non-UGLR	1.72	1.74	1.34	2.12	2.63	24.20	2.64	4.20	59.5
Rural	1.80	1.70	-5.56	1.37	2.07	51.21	1.54	3.07	100.0
Urban	1.78	1.81	1.51	3.34	4.53	35.62	2.86	5.55	94.2
Total	1.79	1.76	-1.68	2.46	3.43	39.51	2.43	4.75	95.4

[a] The change was significantly different at 10%.

Table 7.6 Debt to net worth ratio by region at start-up[a]

	<0		0		0–1		1–3		>3		
	No.	%	No.	%	No.	%	No.	%	No.	%	Ave.[b]
UGLR (n = 75)	1	1.3	16	21.3	14	18.7	15	20.0	29	38.7	4.13
Non-UGLR (n = 37)	0	0.0	11	29.7	6	16.2	9	24.3	11	29.7	3.07
Rural (n = 53)	0	0.0	13	24.5	9	17.0	13	24.5	18	34.0	3.41
Urban (n = 58)	1	1.7	14	24.1	11	19.0	11	19.0	21	36.2	3.16
Total (n = 111)	1	0.9	27	24.1	20	17.9	24	21.4	40	35.7	3.73

[a] There were twenty-two firms with a zero net worth at start-up and information was missing on another firm. They are not included in the average.
[b] There were no significant differences at 10% among regional averages.

Table 7.7 Debt to net worth ratio by region in December 1978[a]

| | <0 | | 0 | | 0-1 | | 1-3 | | >3 | | |
	No.	%	No.	%	No.	%	No.	%	No.	%	Ave.[b]
UGLR (n = 88)	3	3.4	6	6.8	34	38.6	21	23.9	24	27.3	2.66
Non-UGLR (n = 41)	1	2.4	3	7.3	15	36.6	10	24.4	12	29.3	2.57
Rural (n = 59)	2	3.4	4	6.8	26	44.1	15	25.4	12	20.3	1.87
Urban (n = 70)	2	2.9	5	7.2	23	32.9	16	22.9	24	34.3	3.28
Total (n = 129)	4	3.1	9	7.0	49	38.0	31	24.0	36	27.9	2.65

[a] One firm with a zero net worth was not included in the average and information was missing for four others.
[b] There were no significant differences at 10% among regional averages.

ting relatively fewer firms in the non–UGLR borrowed debt capital to begin their businesses compared to firms in the UGLR. The number of firms with a zero debt to net worth ratio decreased dramatically over the period indicating that many firms not borrowing at start-up due to preference or inability were able to borrow debt capital once started. By December 1978 the majority of all firms were borrowing at least their net worth contribution (> 1), indicating debt had become relatively more available as the firm continued operation.

This evidence suggests the firms in the UGLR and rural areas were able to acquire debt capital to a relatively greater extent than firms located outside UGLR and in urban areas. But this preliminary finding may mask other differences.

CAPITAL STRUCTURE

Equity capital

While the common perception is that equity capital is not borrowed, it can be. The owner may borrow his/her equity contribution as a personal loan from a friend, relative, bank, or insurance company. Owners were asked about their 'personal borrowing' of equity. The proportion of firms by type of organization and area with owners who borrowed some of their equity contribution to the business varied among the regions (see Table 7.8). A higher percentage of firms borrowed some equity capital in non–UGLR and urban areas.

There was little variation in the sources of borrowed equity capital across the regions given the type of business organization. Insurance policies were the most likely source of sole proprietors. Insurance

Table 7.8 *Proportion of firms with some borrowed equity contribution by owner(s) by region at start-up*

	Sole proprietor-ship (%)	Partner-ship (%)	Corpo-ration (%)	Total (%)
UGLR	18.2	42.9	62.1	35.2
Non-UGLR	23.8	66.7	52.6	51.7
Rural	19.5	40.0	57.9	35.0
Urban	20.0	60.0	58.6	44.6
Total	19.7	50.0	58.3	40.3

policies and banks were the most likely source for partnerships and corporations. There was considerable variation in sources of borrowed equity capital between sole proprietorships, partnerships, and corporations. Sole proprietors and incorporators were more likely to approach family and friends as a source of borrowed equity and partnerships used insurance policies, etc.

Debt capital

Numerous forms of debt capital are possibilities for the new firm to finance fixed and current assets. The survey examined loans, supplier credit, lease financing, and installment purchases. The type of debt used depends on the types of assets financed, preferences of the owner, and availability of that form of financing. This paper will discuss only formal loans and supplier (dealer/trade) credit.[4]

During the interview, businesses were asked to identify up to six short-term, four medium-term, three long-term, and three real estate loans they made between start-up and December 1978.[5] A total of 413 loans were identified.

Table 7.9 *Average number of loans by type*

	UGLR	Non-UGLR	Rural	Urban	Total
Short-term	1.26	1.33	1.10[a]	1.43[a]	1.29
Medium-term	.69	.65	.77	.61	.68
Long-term	.65	.53	.48[a]	.72[a]	.61
Real estate	.57[a]	.35[a]	.58	.43	.50

[a] Significantly different at 10%.

Table 7.9 presents information on the average number of loans by type of loan. On average, urban and non–UGLR firms held more short-term loans than did rural and UGLR firms. Short-term loans for UGLR firms averaged 35% larger than those for non-UGLR firms (Table 7.10). Short-term loans for urban firms averaged 80% greater than those for rural firms.

The number of medium-term loans held by rural versus urban firms varied more than for UGLR and non-UGLR firms. Urban firms held fewer but larger medium-term loans, while UGLR firms held larger medium-term loans. On average, UGLR firms held more but smaller

Table 7.10 *Average size of original loan*

	UGLR ($)	Non-UGLR ($)	Rural ($)	Urban ($)	Total ($)
Short-term	23,655	17,570	14,505[a]	26,080[a]	21,639
Medium-term	14,614	9,875	10,924	15,438	13,156
Long-term	37,114	38,943	29,157[a]	42,296[a]	37,649
Real estate	53,072	76,760	54,984	62,084	58,375

[a] Significantly different at 10%.

Table 7.11 *Average loan structure of firms*[a]

	Short-term (%)	Medium-term (%)	Long-term (%)	Real estate (%)	Average total loans ($)
UGLR					
Total	31.7	10.7	25.5	32.1	94,422
Rural	20.5	11.4	24.0	44.1	65,177
Urban	39.6	10.5	27.0	22.9	128,764
Non-UGLR					
Total	30.1	8.3	26.9	34.6	77,327
Rural	20.4	13.6	5.3	50.8	100,645
Urban	30.0	6.4	35.0	28.6	71,155
Rural	22.6	11.9	20.0	45.5	70,498
Urban	36.0	9.0	29.2	25.8	103,886
Total	31.2	10.0	25.9	32.8	88,986

[a] Based on cumulative total of loans made.

long-term and real estate loans than did non-UGLR firms. Rural firms
held fewer and smaller long-term and more and smaller real estate loans
compared to urban firms. Data in Table 7.3 indicates that some of these
differences can be associated with differences in the size of firms among
regions. Yet the differences suggest some variation in the functioning
of the local credit market.

Table 7.11 contains data on the average original loan amount the
firms borrowed during the first few years of their existence. It does not
represent a point in time but rather a cumulative total of formal loans

made. Andrews and Eiseman (1981) estimated that in 1975 the capital
structure of all nonfarm, nonfinancial firms was 28.1% short-term and
22.6% long-term loans. The cumulative total reported in Table 7.11 is
remarkably similar with respect to the total for UGLR, non-UGLR,
and the state. Yet when the regions are subdivided, there is considerable
variation. UGLR rural firms used half the short-term loans that UGLR
urban firms did, and twice as much real estate loans. For non-UGLR
firms the differences in the use of short-term loans by rural or urban
firms was not as dramatic, but rural firms apparently had to pledge real
estate for long-term loans to a much greater extent than did urban
firms. Non-UGLR rural firms were able to acquire relatively more
medium-term loans than did urban firms. When examining the rural/
urban firm dichotomy, urban firms relied more on short-term loans
than did rural firms, while the opposite occurs for real estate loans. This
may reflect the preferences of firms or financial institutions in rural
areas in requiring real estate as collateral for loans.

The difference in lending sources among areas was minimal. Banks
were the source of at least 83% of the short-term loans, 82% of the
medium-term loans, and 59% of the long-term loans among all
regions. Previous owners were an important source (21%) of the
long-term loans in rural areas. This may represent a reluctance by
'rural' banks to make long-term loans or a different attitude by the
former owner about financing the sale. The data does not permit
distinguishing the cause. Only for real estate loans did banks contribute
less than half the number of loans reported. The former owner was
almost as likely as the bank to be the source of real estate loans.

The evidence suggests while the firms made similar numbers of
medium-term loans, the average number of short-term, long-term and
real estate loans varied among areas. Furthermore, the size of loans
varied among areas and this was not totally a result of firm size. Finally,
commercial banks were easily the most common debt capital source,
particularly in rural areas.

Supplier credit

All but four of the 134 firms used supplier credit in 1977, and only two
of the firms did not use supplier credit in 1978. The 1977 average
monthly supplier credit did not vary dramatically between geographic
areas (see Table 7.12). However, the much higher rate of growth in
supplier credit for firms in the UGLR reversed the relative position
with firms outside the UGLR between 1977 and 1978. During the same

Table 7.12 *Average monthly supplier credit by region*[a]

	1977 ($)		1978 ($)		% change
UGLRC	24,758	(n = 89)	40,818	(n = 90)	64.8
Non-UGLRC	28,685	(n = 41)	32,706	(n = 42)	14.0
Rural	27,696	(n = 58)	46,579	(n = 59)	68.0
Urban	24,628	(n = 72)	31,573	(n = 73)	28.2
Total	25,997	(n = 130)	38,217	(n = 132)	47.0

[a] There was no significant difference at 10%.

time, the higher growth rate of supplier credit for rural firms substantially widened the difference with urban firms.

Most of the firms operating without supplier credit did so because they could not qualify for it as a result of their short or otherwise unacceptable financial and operating history, or because they found it unnecessary or inconvenient. Acquiring some supplier credit at start-up was not considered much of a problem by most of those interviewed. Many of the respondents said that as new accounts they were watched very closely by their suppliers for the first six months. There were many cases, especially among the grocery retailers and eating and drinking establishments, where suppliers required cash on delivery or even cash with the order regardless of the age of the business.

The average monthly amount of supplier credit increased over 26% for all firms and as much as 61.4% for local and long distance trucking firms, 36.0% for general construction contractors and 34.3% for durable wholesale trade (see Table 7.13).

While credit terms were not standardized among SIC groups, the maturities offered, availability of early payment discounts, and interest rates charged were similar within each SIC group. Clearly, firms in a given line of business (same SIC) purchase from many of the same suppliers causing much of the uniformity among similar business types.

Within some SIC groups there is no variation in supplier credit terms; in others there is great variation. Supplier credit terms for the retail liquor stores are set by law and are the same for all in that group (for purchases of beer, wine, and liquor at least). On the other hand, for the equipment dealers the variability of credit terms related more to the type of equipment purchased rather than to the purchaser. For example,

Table 7.13 *Monthly average supplier credit by type of business*

SIC	Number of firms reporting	Average 1977 ($)	1978 ($)
15 Contract construction	14	7,775	10,577
3599 Machinery manufacturing (except electrical)	16	14,505	16,470
421 Trucking, local and long distance	16	2,371	3,827
508 Wholesale trade, durable goods	17	76,579	102,866
514 Wholesale trade, nondurable goods	7	50,014	63,029
541 Retail food stores	16	47,929	58,220
571 Retail furniture and home furnishings	12	18,042	19,625
58 Eating and drinking places	20	3,815	4,723
592 Liquor stores	13	13,192	14,615
Total	131	24,323	30,755

a single purchaser may receive different terms for each different type of equipment purchased and all purchasers will receive similar terms on the same equipment.

While the terms offered on a given type of product or by a given supplier were similar for all credit purchasers, use of the terms was not uniform among the respondents. Most of the respondents (72%) took maximum advantage of the supplier credit terms offered. However, a few of the respondents reported they routinely paid off their supplier credit balances on some weekly or monthly schedule even if a longer maturity was available either free of charge or at an interest rate below what they would have to pay for borrowed capital. In these cases, the credit balances were small enough that the economic gains available were minimal compared to the effort needed to take full advantage of the credit.

Almost three out of five (56.3%) of the respondents reported having been offered early payment discounts by suppliers. They took advantage of those offers 64.4% of the time. A lack of sufficient operating capital was by far the most common reason for not always taking advantage of offered discounts.

A comparison of the average firm in Tables 7.3 and 7.12 yields some insights about the role and importance of supplier credit to these businesses. Recognizing that start-up and 1977 are not completely comparable, average monthly supplier credit declined as a proportion of total debt between start-up and December 1978 even though the use of supplier credit increased. Suppliers, because of their relationships to the new small business, provide a very important financing substitute to more formal financial institutions. However, as the small business grows it gains access to other capital sources and supplier credit becomes relatively less important.

CREDIT DENIAL

Whenever a business approaches a lender there is always the possibility that particular credit application will be denied. The reason for the denial and the firm's response are useful for understanding the credit market facing new small businesses.

The businesses interviewed were asked if the firm had had a loan application denied between start-up and December 1978. Thirty-three firms (24.7%) responded they had been denied credit.[6] Of these thirty-three firms, only eight were unable to acquire credit eventually for that project. The remaining twenty-five firms received some other type of debt financing for the project originally denied.

There was little difference in denial rates between the urban (24.3%) and rural (25.0%) areas. Firms located in the UGLR, however, experienced a 50% higher rate of credit denial than those located outside the UGLR, e.g., 27.5% versus 18.6%. There was little difference in credit denial experience for UGLR firms when further classified rural/urban (29.4% and 25.0% respectively) but no rural firms in the non-UGLR experienced a credit denial. The data does not permit determining if this is due to differences in the quality of credit application among areas or differences in lending attitudes by financial institutions.

Since the firms surveyed in the UGLR and urban areas were larger, it is appropriate to ask if credit denial is related to firm size. The relationship between firm asset size and credit denial varied among regions. The start-up total assets of firms denied credit are larger than the area average in the non-UGLR and urban locations. The opposite is true for UGLR and rural firms. Between start-up and December 1978 the total assets of firms denied credit exhibited a faster rate of growth than the area average in all areas except urban areas.

Nearly half the firms with December 1978 total assets of more than $100,000 experienced credit denial. This suggests that the loan requests may have pushed the individual loan limit of the lender. While the quality of credit application cannot be judged, this offers tentative evidence that larger credit requests may experience more credit acquisition difficulties. The firms with total assets of less than $40,000, both at start-up and December 1978, and being denied credit, decreased by 50% over the period. Thus, the smaller firms, once established, found access to credit easier.

The presence of assets, e.g., collateral, does not necessarily indicate that the firm has the cash flow to pay off the loan.[7] Volume of sales is a proxy for cash flow. The data indicates that the firms denied credit had greater sales than the average firm in each geographic area, suggesting that the credit denial was not based on limited cash flow as indicated by gross sales.

The inability of the firm to pledge 'free and clear' assets for the current loan request may be one reason for credit denial. Of the firms denied credit, 42% had a net worth of $10,000 or less at start-up. Of the firms denied credit, 21% had at best a zero net worth at start-up. None of the firms with at least $50,000 of net worth at start-up were denied credit. The general tone of this evidence suggests that the firm's inability to produce 'free and clear' assets or net worth was a major influence on initial credit denial.

Statewide, firms denied credit had less net worth both at start-up and December 1978, but grew faster than all firms. This same phenomenon occurred for firms denied credit in the UGLR and rural areas. In all regions, the firms denied credit had a faster rate of growth in net worth than did the average firm.

In all regions, the firms denied credit indicated the refusal of the lending institution to accept the proposed use of funds as the predominant reason for denying credit. This phenomenon may occur because the project was infeasible[8] or the firm was unable to convince the lending institution that the project was feasible; or it might result from ignorance on the part of the lending institution regarding the type of project proposed. The unacceptability of the use of funds suggests that lenders were exerting a large degree of control in the firm's operations.

A higher percentage of firms in the urban and non-UGLR regions were refused credit because of the lack of equity, and a higher percentage of firms in the rural and UGLR were refused because of the lack of collateral. However, urban and non-UGLR firms denied credit averaged more net worth at start-up and in December 1978, and grew faster than did all firms in the respective region. Rural and UGLR firms denied credit averaged less total assets at start-up, but grew much faster than did all firms to exceed the all firm average for their respective region by December 1978. Finally, a relatively higher percentage of firms in the UGLR and rural areas were refused credit because they were viewed as an unacceptable credit risk.

In some cases, reasons for credit denial appear reasonable, but the 75% success rate on approaching another similar lender reinforces the need for firms not to accept a refusal as the final verdict. The end result of these comparisons is that lending criteria vary among regions and banks. The decision making capacity of individual banks appears to be a significant force affecting the flow of capital within an area.

CAPITAL STRESS

One of the major objectives of this study was to determine the extent of capital availability for new small businesses. One line of theory suggests that capital will be available if the user is willing to bid it away from other users. This has a certain appeal, but implicitly places newer and smaller businesses at a disadvantage to the more established and larger firms which are in the bidding for capital and which have greater ability to pay. The higher rates of growth for the firms denied credit indicates that they use credit effectively once acquired.

To determine the capital availability for new small firms, firms in the sample were asked if: (1) they perceived their banks as having met their credit needs; (2) they perceived all financial institutions as having met their credit needs; and (3) they would be able to find capital (debt or equity) within thirty miles for expansion of their business, regardless of expansion plans. The latter question required a direct yes or no answer. The 'needs' questions required the firm to assess whether their needs had been met 'very well,' 'well,' 'fair,' 'poor,' and 'very poor.'

Tables 7.14 and 7.15 show the regional opinions about how well credit needs had been met. It is obvious that relatively more of the firms in the UGLR and rural regions held fair, poor, and very poor opinions as opposed to firms in the non-UGLR and urban regions. The capital markets in the UGLR and rural areas of the state appeared to be functioning less well in meeting the capital needs of new small businesses than were capital markets elsewhere. This difference may arise from credit evaluation judgments, size of financial institution, or even access to alternative credit sources.

Table 7.16 presents information on the proportion of firms believing that they could not acquire adequate capital (debt or equity) for expansion within thirty miles of their present location. The perception of unavailability of capital within thirty miles was much more pronounced for firms located in the UGLR and rural areas. The firms, by

Table 7.14 *How well have overall credit needs been met by financial institutions, by region*[a]

	Very well		Well		Fair		Poor		Very poor	
	No.	%	No.	%	No.	%	No.	%	No.	%
UGLR (n = 91)	51	56.0	26	28.6	11	12.0	–	–	3	3.3
Non-UGLR (n = 42)	27	64.3	12	31.0	–	–	–	–	2	4.8
Rural (n = 60)	34	56.7	17	28.3	8	13.3	–	–	1	1.7
Urban (n =73)	44	60.3	22	30.1	3	4.1	–	–	4	5.5
Total # (n = 133)	78	58.6	39	29.3	11	8.3	–	–	5	3.8

[a] One firm did not respond.

Table 7.15 *How well has local bank met your credit needs, by region*[a]

	Very well		Well		Fair		Poor		Very poor	
	No	%	No.	%	No.	%	No.	%	No.	%
UGLR (n = 88)	53	60.2	14	15.9	10	17.2	6	10.3	5	8.6
Non-UGLR (n = 38)	26	68.4	6	15.8	4	10.5	–	–	2	5.3
Rural (n = 58)	34	58.6	9	15.5	8	13.8	5	8.6	2	3.4
Urban (n = 68)	45	66.2	11	16.2	6	8.8	1	1.4	5	7.4
Total # (n = 126)	79	62.7	20	15.9	14	11.1	6	4.8	7	5.6

[a] Eight firms did not respond.

Table 7.16 *Unable to find capital within thirty miles of region*

	Number of firms	% of region
UGLR (n = 91)	25	27.5
Non-UGLR (n = 43)	5	11.6
Rural (n = 60)	15	25.0
Urban (n = 74)	15	20.3
Total (n = 134)	30	22.4

region, believed they could raise the following average amounts of capital: UGLR – $38,858; non-UGLR – $33,954; rural – $22,934; and urban – $48,919. Thus, there is no apparent relationship between size of capital needs and belief about capital availability. Rather, it appears to be a function of density of settlement patterns, e.g., availability of alternative financial institutions.

A variable was created to capture the various dimensions of capital stress. Firms were perceived as experiencing capital stress if they: (1) were denied credit at any time; (2) felt their banks did not meet their needs (fair, poor, or very poor); (3) felt all financial institutions did not meet their needs (fair, poor, very poor); or (4) could not locate capital within thirty miles of their operations. A positive response to any of these questions placed the firm in the stress category.

The number of firms who perceived some type of capital stress are shown in Table 7.17. Relatively more firms in the UGLR area perceived stress than did firms in the non-UGLR area, and relatively more firms in the rural areas perceived stress than did firms in the urban areas. When UGLR and non-UGLR are further divided into rural/urban, rural firms in the UGLR and urban firms in non-UGLR experienced more capital stress. This indicates that new small businesses in these areas believe that capital markets were functioning less well than in other areas of the state.

CONCLUSIONS

This study of the capital structure of new small businesses and their difficulties in acquiring capital for start-up and during the first few years of operation suggests that capital markets in Wisconsin appear to be

Table 7.17 *Businesses experiencing capital stress by region*

	Number of firms	% of region
UGLR (n = 91)	46	50.5
Rural (n = 51)	28	54.9
Urban (n = 40)	19	47.5
Non-UGLR (n = 43)	13	30.2
Rural (n = 9)	2	22.2
Urban (n = 34)	11	32.4
Rural (n = 60)	29	48.3
Urban (n = 74)	30	40.5
Total (n = 134)	59	44.0

functioning relatively well for new small businesses. But this conclusion must be conditioned by four caveats. First, the data is based on firms and entrepreneurs who successfully started and continued operating up to four years at the time of the interview. The data provides no insight about how well the capital market is functioning for people and firms becoming discouraged and not trying to start or failing once started. Second, the success rate of persistence suggests that the decision making guidelines and capacity of lending institutions is highly variable and may be as much an impediment to well functioning capital markets as any other factor. Third, the consolidation of various dimensions of capital market functioning into a capital stress index indicates almost half (44%) of the firms did not believe capital markets were functioning adequately. Fourth, there is wide regional variation in capital stress. Capital markets appear to function less well in areas over thirty miles from cities of at least 30,000 people (rural) and in the sparsely settled, lower economic activity area of the UGLR.

The purpose of this study was to examine the functioning of the capital markets for new small firms in different geographic areas of the State of Wisconsin. There were four general hypotheses derived from modern capital market theory. First, the capital market will make appropriate allocations of capital over space and among uses. The evidence is mixed. The capital structure of the various types of businesses and among the various areas was different as reflected in debt to net worth ratios, number and size of loans, use of supplier credit, and informal equity capital sources. This implies that the market was able to accommodate the differences among businesses and regions. However, the rates of credit denial and level of capital stress varied substantially among some geographic areas. Second, the decision making capacity of financial institutions would not hinder the movement of capital to appropriate uses. The firms denied credit grew faster than the average of all firms in their respective area and the reasons cited for the denial often did not correlate with the financial conditions of the firms. The failure of financial institutions to make positive responses to lending requests, eventually funded by other institutions, suggests that the decision making capacity of financial institutions may be a major constraint in the functioning of the capital markets regardless of area. Third, a firm's access to the capital market would improve with operating experience or record. The amount of debt capital used increased with length of time the firm operated, but the use of debt relative to net worth fell. There was a decline in dependence on informal capital sources, e.g., family, friends, suppliers, as the firm became more established. Fourth, size and rate of growth of the firm would not affect access to capital. Within the limited variation of firm sizes examined in this study, the evidence is mixed about size and access to debt capital. However, firms denied credit grew faster than the area average, suggesting that the loan evaluation capacity of lenders may constrain regional economic growth.

NOTES

1 The data used in this article was collected through the support of Technical Assistance Grants 10020462 and 10820408 from the Upper Great Lakes Regional Commission, Hatch Project No. 2240 and EDD/ERS/USDA project Q550.
2 Total employment is the sum of owners and unpaid family labor, full-time workers, and part-time workers.
3 Note the data in Tables 7.6 and 7.7 is not strictly comparable with Table 7.3.
4 The reader interested in a fuller discussion of other sources of debt is encouraged to read *Regional Variation in Capital Structure of New Small Businesses in Wisconsin*, Res. Bul. R3209, Univ. of Wisc. College of Agriculture and Life Sciences, January 1983.
5 For purposes of this study a short-term loan was defined as a loan having maturity of 12 months or less. A medium-term loan had a maturity of 13 to 36 months. A long-term loan had a maturity of at least 37 months and was not for real estate purposes.
6 The Council for Northeast Economic Action found nationwide 25% of the firms interviewed were unable to acquire intermediate or long-term funds (CNEA, 1981).

7 There are two general forms of lenders. They are asset lenders and cash flow lenders. The latter is more likely to support new small businesses.
8 Indirect evidence refuting this is that twenty-five of the thirty-three firms had the project financed by another lender and the firms denied credit grew faster than the average for firms not denied credit.

REFERENCES

Andrews, Victor L. and Peter C. Eiseman, *Who Finances Small Business Circa 1980?*, staff paper for The Interagency Task Force on Small Business Finance, Washington DC: Small Business Administration (Nov. 1981).

Armington, Catherine and Marjorie Odle, 'Small Business – How Many Jobs?', *Brookings Review* (Winter 1982).

Birch, David, *The Job Generation Process*, Washington DC: Economic Development Administration (1979).

Council for Northeast Economic Action, *An Empirical Analysis of Unmet Demand in Domestic Capital Markets in Five US Regions*, Washington DC: Economic Development Administration (Feb. 1981).

Daniels, Belden Hull and Harry Lirtzman, *Providing Capital to Small Businesses. An Evaluation of State Development Incentives*, Washington DC: Small Business Administration (Nov. 1980).

Dunkelberg, William C. and Jonathan A. Scott, 'Problems, Policy Issues and Research Needs in Rural Financial Markets: The View of Small Business', in *Rural Financial Markets: Research Issues for the 1980's*, Chicago, Ill.: Federal Reserve Bank of Chicago (Aug. 1983).

Mikesell, James and Steve Davidson, 'Financing Rural America: A Public Policy and Research Perspective', in *Rural Financial Markets: Research Issues for the 1980's*, Chicago, Ill.: Federal Reserve Bank of Chicago (Aug. 1983).

Small Business Administration, *The State of Small Business*, Washington DC: US Government Printing Office (March 1983).

Shapero, Albert, *The Role of the Financial Institutions of a Community in the Formation, Effectiveness and Expansion of Innovating Companies*, Washington DC: Small Business Administration (Feb. 1983).

White, Lawrence J., *Role of Small Business in the American Economy*, staff paper for The Interagency Task Force on Small Business Finance, Washington DC: Small Business Administration (Sept. 1980).

8

The world of small business: turbulence and survival

ANN R. MARKUSEN and MICHAEL B. TEITZ[1]

ANN R. MARKUSEN and MICHAEL B. TEITZ[1]

SMALL BUSINESS AS REGIONAL EMPLOYMENT GENERATOR

Traditionally, the engine of local and regional economic development and employment growth has been perceived to be investment in new or expanded large-scale industrial or commercial facilities. Indeed, that is the basis upon which the greater part of local development efforts still rely. In recent years, however, an alternative view has been put forward and widely accepted. This view sees small business as a key element in economic development, and, especially, in the generation of new employment. While not without its critics, this conception of small business as a principal source of dynamism in local and regional economies suggests the need for careful assessment of the role and nature of small business. Such an assessment has thus far been based largely on statistical analysis of the employment-generating effects of small business. The intent of this paper is to complement that work with a grounded picture of small business as revealed in case interviews with a limited number of firms. But before discussing the results of the survey, it is useful to recapitulate the status of investigations into the role of small business in the aggregate.

In the United States, research on the employment effects of small business is largely associated with the work of David Birch. Birch's initial study (1979) was based on a very large data file of firms collected by Dun and Bradstreet, a credit rating organization. According to his estimate, firms with twenty or fewer employees accounted for 66 percent of *net* employment growth in the United States between 1969 and 1976. Subsequent work has both supported and challenged Birch's findings. Teitz *et al.* (1981) analyzed an unemployment insurance data set representing virtually all firms in California for the period 1975–9.

They found that firms in this size category produced some 56 percent of net new jobs in the state. More importantly, this work suggested that the dynamics of employment generation were such that much of that total may be due to a relatively small number of fast-growing firms. Birch's work has been criticized on several grounds. Armington and Odle (1982) have also analyzed the Dun and Bradstreet data, but come to rather different conclusions. They argue that the large proportion of employment generated by small business may be due to the use of establishment data that suppress the connection of small units to larger parent firms. This debate is not yet resolved. A further question of the quality and stability of jobs created in small firms is raised by Gordon (1979) and Bluestone and Harrison (1980), all of whom assert that the higher attrition experienced by small firms dooms most of the jobs that they create to be short-term. This characteristic has long been noticed in studies of the dynamics of populations of firms, for example Ijiri and Simon (1977). In addition, small businesses may exhibit lower produc- tivity and pay lower wages than large firms. Fothergill and Gudgin (1979) question whether Birch's results depend on the growth of the service sector in the United States. Their analysis of manufacturing firms in the British East Midlands region indicated only a modest contribution by small businesses. Teitz *et al.* (1981) also showed a lower rate of job generation in manufacturing by small firms.

Despite these criticisms, the idea that small business is important to economic development and employment growth is now well estab- lished. Local and regional agencies are incorporating small business development explicitly into their development strategies. Research on small business behavior and its contribution to development and employment is continuing. At the same time, aggregate analysis alone is insufficient for the development of policy. If small businesses are to be fostered and encouraged, we need to know more about the way in which they are founded, survive, grow, and fail. Such understanding comes from analysis of the reality in which these firms live and function. Adequate description of the reality of small business is a difficult task. Small firms are vast in number and diverse in nature and function. Our objective is to explore their world by means of inter- views with a small sample of firms. This study is not a data gathering and hypothesis testing enterprise. Rather, it is intended to reveal some of the richness and complexity of the small business environment. From this process, we may gain a better insight into the situations in which small firms actually operate, and perhaps begin to understand how they behave.

THE SURVEY AND INTERVIEWS

The small business environment was investigated by interviewing a sample of twenty-eight Bay Area small businesses, chosen at random and covering a range of employment size categories and sectors. Unlike many other studies, the sample was not restricted to the manufacturing sector; rather, representative firms were chosen from all sectors, including wholesaling, retailing, business and personal services, construction, transportation, and finance, insurance and real estate (FIRE).[2] Within the size and sectoral categories, firms were selected from the SIC index edition of the *Contacts Influential Commerce & Industry Directory: East Bay Edition* (Influential Contacts, Ltd, 1979).[3] No attempt was made to control for geographic location within the East Bay Area.

The actual selection of the firms from the directory was statistically random. Table 8.1 shows the composition of the combined sample from the pre-test and the main survey. The composition by sector and by size category of the final sample is shown in Table 8.2. The firms are fairly evenly distributed throughout the size range. Table 8.3 indicates how the sectoral distribution of firms in the sample compares with the statewide distribution. It should be noted that the small number of cases in each sector and size category cautions against generalizing along either of these dimensions. The point, rather, was to ensure coverage of all types of small businesses.

Table 8.1 *Structure of the initial total sample including pre-test*

Sector	Frequency	%	Number of employees		
			1–5	6–49	50–100
Manufacturing	15	23.8	5	6	4
Construction	6	9.5	2	2	2
Transportation	3	4.8	1	1	1
Wholesale	6	9.5	2	2	2
Retail	14	22.2	5	6	3
Services	13	20.6	4	6	3
FIRE	6	9.5	2	2	2
Total[a]	63	100.0	21	25	17

Notes
Based on information in the *Contacts Influential Directory*.
[a] Total may not add to 100 per cent owing to rounding.

Table 8.2 *Final sample structure by size and sector*

| Sector[a] | Frequency | % | Number of employees | | |
			1–5	6–49	50–100
Manufacturing	5	1	3	1	
Construction	2	1		1	
Transportation	2	1	1		
Wholesale	1				
Retail	8	3	3	2	
Services	5	1	1	3	
FIRE	2	1		1	
Total[b]	28	10	9	9	
Percentage response	44				

Notes
[a] Based on actual figures obtained from the firms during interviews.
[b] One with more than 100 employees.

Table 8.3 *Comparison of percentage distributions of firms in sample and statewide*

Sector	Final sample (firms) (%)	Statewide[a] (reporting units) (%)
Manufacturing	17.0	9.2
Construction	7.1	10.4
Transportation	7.1	3.4
Wholesale	14.3	8.2
Retail	28.6	24.4
Services	17.9	35.4
FIRE	7.1	8.9
Total	100.0[b] N = 28	100.0[b] N = 419,288

Notes
[a] Adjusted to exclude other sectors from the total. Derived from the California Employment Development Department, 524 Report.
[b] Percentages may not add to 100 owing to rounding.

The survey itself probed small business experience first-hand to find out what conditions were affecting the business in both the short- and long-run, and to generate new hypotheses about the problems and prospects of small business. Unlike most small business surveys, which deal with only one or two issues, our objective was to elicit a broad

Table 8.4 *Characteristics of sample firms*

Firm, by industry	Number of employees	Gross annual sales	Age of firm (years)	Ownership status
A. Manufacturing				
Aerial security systems	40	$3m.	52	Closely held corp.
Container products	34	$3m.	5	Corporation
Fiber production	85	$800,000	18	Corporation
Recreation equipment	3	$250,000	17	Partnership
Wood products	7	$1.2–1.5m.	87	Closely held corp.
B. Construction				
Excavation	5	$164,000	16	Partnership
Underground utilities	85	$2–8m.	30	Corporation
C. Transportation				
Air transportation	65	$1–2.5m.	6	Corporation
Emergency transportation	40	$1m.	33	Closely held corp.
D. Wholesale				
Book publisher/distributor	2	$1.4m.	6	Corporation
Building supplies	6	$1.8m.	15	Corporation
Furniture importer	2	$150–250,000	31	Proprietor
Machine tool distributor	2	$300,000	11	Proprietor
E. Retail				
Florist	6	$275,000	20	Closely held corp.
Food market (a)	4	$500,000	18	Proprietor
Food market (b)	85	$11m.	50	Closely held corp.
Food market (c)	100	$15m.	12	Closely held corp.
Liquor store	2	$175,000	33	Proprietor
Paint store	3	$400,000	25	Closely held corp.
Recreational vehicles	22–5	$3m.	12	Closely held corp.
Restaurant	8	$140,000	15	Partnership
F. Services				
Advertising firm	4	$450,000	26	Corporation
Auto repair	165	$6m.	38	Corporation
Dry cleaning	7	N/A	5	Closely held corp.
Health maintenance organization	70	N/A	10	Non-profit
Marketing research	55	$350,000	4	Partnership
G. Finance, insurance, real estate				
Realtor	6	N/A	13	Proprietorship
Securities investment	65	$7.5m.	N/A	Partnership

Number of firms in sample = 28.

range of experiences in the operation of small businesses. Interview questions were therefore structured to develop both qualitative and quantitative responses on a broad spectrum of topics, including the business environment, entrepreneurship, finance capital, labor, location and physical facilities, other private sector factors, and the effect of the public sector, especially taxes and regulation.[4] Table 8.4 summarizes the general characteristics of the firms surveyed.

THE COMPETITIVE ENVIRONMENT

The survey sought first to gain a picture of the conditions influencing the operation and development of the small business in the sample. Informants were asked about the underlying dynamics affecting their businesses, the role played by competition, strategies employed to overcome adverse dynamic and competitive factors, and attitudes toward growth and innovation.

Underlying dynamics

Small businesses are often depicted as being either very risky and short-lived or else stable, permanent residents of the local economy. Our survey turned up evidence of a dynamic and turbulent environment. The longevity of businesses included in the sample was quite high; the mean age was nearly twenty-two years, while the median age was eighteen. However, most of the businesses were experiencing some kind of change. While eight were stagnant or declining, the rest were facing drastic crises (six), growing steadily (five) or expanding rapidly (eleven). Thus, the typical small business seemed to be facing some degree of uncertainty. Those who escaped tended to be local retailers in stable markets. Although the specific causes of change and stability were often complex, three types of factors appeared to be involved: cyclical, sectoral, or spatial.

Cyclical factors created problems for fourteen, or half of the firms. Five firms faced seasonal fluctuations in demand or supply conditions, causing severe although not necessarily fatal cash flow problems. Another firm was subject to agricultural cycles, which have at times lasted more than one season. By far the greatest number of cyclically induced problems, however, originated in business cycle fluctuations and other general economic trends. Both recession and inflation posed major obstacles for these small businesses.

Recession operates on both the demand and supply sides of the small business economy. The 1981 recession affected particularly those firms involved in construction, in wholesaling, and in supplying consumer services (including the restaurant) and retail goods (for example, the furniture supplier and the dealers in recreational vehicles and aviation equipment). Some firms were severely squeezed for cash as their customers stretched out their payments and accounts receivable rose. On the supply side, some firms found that their suppliers passed on the brunt of the recession to them – an aviation-related firm, for instance, had to absorb the manufacturer's excess output. A dealer in power tools reported that a big company with whom he competes gave unmatchable 40 percent trade-in discounts to keep its volume up. Both types of problems reflect the small firm's vulnerability to tied-buying arrangements or limited competition on the supply side.

The cyclical problem was compounded by inflation. As interest rates rose to reflect both inflation and uncertainty about future price increases, small business customers faced greater difficulties in financing purchases from wholesalers and durable goods dealers. Many encountered difficulties in repaying short-term borrowings. The aviation-related firm, for instance, seemed to be in a position where it could neither sell nor borrow to meet its debts. Rising costs had other effects too. A restaurateur noted that inflation lowered his volume and altered the clientele from families to young professionals as higher menu prices led to fewer meals served. Moreover, as one small manufacturer argued, inflation tended to break down traditional dealings between buyers and sellers. Previously, he pointed out, small businesses would resist, negotiate, and haggle with suppliers and buyers over prices. With permanent inflation, increases are just passed on automatically.

Sectoral causes – the second major group of factors underlying instability – were specific to each industry and were frequently related to product-cycle developments. For instance, a small furniture dealer selling imported Norwegian furniture had at one time distributed throughout the country and employed more than twenty workers. By the late 1970s, he had retrenched to a three-person operation. The rise and fall of his business can be traced to the fashionableness and innovativeness of this style of furniture in the 1950s and 1960s and its imitation, market saturation, and subsequent decline in more recent times. A small manufacturer/dealer of car-top campers and a recreational vehicle dealer experienced similar, although less dramatic, paths. In other cases, decline resulted from innovations that rendered a product or service obsolete. On the supply side, a small manufacturing firm found itself in trouble when its raw material, a byproduct of another industry, disappeared when the latter adopted a new production process. Other sectoral changes that frequently damaged individual small business prospects arose from growing degrees of monopolization or intensification of competition in a market. The small grocery stores, for example, had to struggle much harder than previously to survive the onslaught of the large chain supermarket.

Spatial causes of business instability – the third group of factors – were most frequently attributable to neighborhood changes which worked to eliminate a local clientele or to increase business costs. One of the grocery stores in the survey had fought to stay alive even though its important customers had moved out of the area. In at least two cases, government-sponsored redevelopment projects exacerbated existing businesses' problems. Gentrification pressures that pushed up rents threaten small retailers, while higher crime rates in deteriorating neighborhoods raise insurance and security costs for others.

COMPETITIVE CONDITIONS

Traditional economic analysis often assumes that a small firm faces a competitive market of many other similarly small firms. In this situation, no single firm dominates, price competition prevails, and sellers are generally unaware of each other's behavior. Our cases, however, revealed a strikingly different picture. Most of the firms in the sample operated in markets with a substantial degree of oligopoly or oligopsony in one or more of three forms. First, competition was often restricted among and between the firms directly competing in a product or service line. Second, competition was often restricted among the buyers of the goods and services the small businesses produced. Finally, competition was often restricted in the market for key supply factors such as materials, labor, building space, capital, and wholesale products for retail.

In their own markets, the firm faced varying degrees of competition from rivals. At least half of the firms responded that their competitors numbered fewer than half a dozen. Another third characterized their markets as dominated by a few large firms with many small ones like themselves operating (or trying to) in between. The degree of competition in a market seemed to be related to the stage of the product cycle, to the existence of economies of scale, and to the size of the market. Spatially limited markets are an important cause of a small number of competitors in some sectors. Over 50 percent of the firms sampled operated in markets smaller than the Bay Area and only five had a market more extensive than northern California. Many firms who might be assumed to operate in a highly competitive market (small groceries, a paint store, a dry cleaners, a liquor store) actually had few, if any, competitors because their markets were spatially circumscribed. The geographical size of their markets, however (i.e. the boundaries between their markets and competitors'), seemed to correspond roughly to economies of scale in Loschian fashion.

Since firms often operated in markets with few rivals, they tended to know who their competitors were, and were not simple price takers. Several remarked that their competitors were friendly, while others reported calling competitors when looking for skilled workers. It appeared that information on supplies, prices, and products was shared as well – the construction firm, for example, shared perceptions of products, suppliers, and equipment with its rivals. Firms easily placed themselves within the pecking order in their field and responded keenly and knowledgeably to questions about competitors. Not surprisingly, firms low on the 'totem pole' complained that their poor performance was due to the larger size of their competitors, while higher ranking firms prided themselves on their superior knowledge, skill, and hard work. Only one firm, an advertiser, acknowledged that its laggard performance was due to inability to match a competitor's 'presentation techniques'.

The existence of both large and small firms in a single market seemed to indicate either product cycle evolution with growing concentration or a process of product differentiation, which is described below. While the emergence of large firms dominating a market could be expected to spell problems for smaller firms, one small grocer argued that he remains 'more knowledgeable than big chain managers and more maneuverable' and was thus able to compete successfully. Product and service differentiation continues to account for the survival of small firms.

Many firms faced market power on the part of their suppliers or customers. Seven firms complained of problems arising from dependence on one or a few suppliers. A plumbing wholesaler complained that the non-price competition of his major supplier resulted in too-rapid style changes which frequently rendered his stock obsolete. A wood products firm suffered from shortages of materials available from big suppliers who, being vertically integrated, supplied themselves first during booms. The recreational vehicles dealer protested that his supplier did not innovate sufficiently to keep him competitive. The aviation-related firm resented the local airport authority's monopoly on parking space. The book dealer and the furniture dealer, both relying on limited international suppliers, complained of problems in obtaining consistent, timely, and good quality products.

Six firms complained of dependence on too few customers. In some cases, these were also relatively small businesses. One customer, ironically, was a local Chamber of Commerce, which required the firm to locate in its jurisdiction in order to get its business, but subsequently stopped purchasing the service. In other cases, customers were large corporations or the government. In each instance, some aspect of unequal market power was severely affecting the small business. Third party behavior could also threaten demand; for example, bank policies on lending could prevent customers from financing a purchase.

The lack of stability across the sample and the high incidence of crisis, retrenchment, or expansion cannot be ascribed solely to the business and competitive environment. Some of the problems are due to internal factors, such as management and labor practices, as we discuss below. However, the competitive conditions do seem to explain much of the success or troubles of the sample firms. Most firms carefully watch their competitors, clients, and suppliers across markets. Many are, indeed, dependent on the specific behavior of one or several of these. Although in a few instances, firms reported that contracts with large buyers helped them ride out the recession, and while supplier credit (discussed below) was frequently a prerequisite to starting up or continuing, vulnerability to a few economic actors appeared to be a major fact and problem of small business life.

Small business strategies

We detected a consistent, although not universal, set of strategies that permit small firms to weather their cyclical, sectoral, spatial, and competitive problems. Firms that seemed able to survive employed one of three paths, depending upon the sector they were in and the sources of their problems. We have dubbed these paths the *branching*, *product diversification*, and *product differentiation* strategies.

Eight of the firms, or 28 percent, had or have had branches. Most could be classified as retail or consumer service-oriented – two groceries, a paint store, a dry cleaners, the recreation vehicle dealer, the health maintenance organization, the furniture dealer, and an auto engine rebuilder. All but one had four or less outlets, and two had failed or closed down at the first branching site. Branching appears to be a way for the small business experiencing neighborhood change to soften the blows of market disruption, and a way for expansion-minded entrepreneurs to overcome the growth limits imposed by a spatially-saturated market.

Product diversification was identified by ten firms, or 35 percent of the sample, as their strategy for cushioning themselves against cyclical, sectoral, and oligopolistic conditions. This strategy included operating at different stages of product/service provision. For instance, the book firm both published its own material and distributed material for other publishers, while the trucking firm diversified into the rental business. Firms that undertook product diversification included four manufacturers, two wholesalers, and two durable goods dealers. Diversification appeared to be more common with producers and distributors than with retailers and was generally modest, in that the diversified product was closely related to the existing product and required similar business techniques and expertise. As opposed to the conglomerate pattern of diversifying into often totally unconnected products or services, the small businesses surveyed tended to keep within, and rely upon, their own specific areas of skill. Diversification was especially important to the small manufacturers as a way of riding out changes in the product cycle or coping with long-term structural changes that threaten to render their products obsolete or dry up their material sources. For distributors or durable goods sellers, diversification helped counter recessionary ups and downs, cover cash flow problems, or secure a foothold in a market.

A third strategy frequently mentioned by the firms was product differentiation. This strategy was used by firms across all the sectors that operated in markets containing large-scale or vertically-integrated firms. Small firms stayed competitive in these markets by exploiting specialized market niches, providing superior service, or servicing spatially-neglected markets. All of the grocery stores, for instance, mentioned some degree of product differentiation, such as personalized service, locational convenience, or superior quality meats. A carton manufacturer and the car-top camper manufacturer thrived on customized service. The advertising firm exploited a specialized agricultural/mining market.

Entrepreneurial attitudes toward growth and innovation

Small firm entrepreneurs exhibited a surprising array of attitudes about the desirability of future firm growth. Because an entrepreneur's plans are frequently shaped by personal beliefs or perceptions about what is possible, generalization is difficult. Aspects of the business environment analyzed above may discourage some from trying to expand when they would otherwise like to – a parallel to the 'discouraged worker' notion in employment studies. Since our interviews took place during a moderately strong recessionary period, this effect may have been more marked than usual. But of those responding to questions about future plans, approximately one third both wanted to expand and felt it likely, another third would like to grow but felt it unlikely, another third had no desire to grow. The 'optimistic growers' spanned all sectors, suggesting that even in adverse market conditions a growth strategy is possible. For instance, among the three groceries, one was a satisfied non-grower, one was a discouraged grower (even though he had tried both diversity and product differentiation, his efforts could not overcome the almost total erosion of his neighborhood market), and one was an optimistic grower.

Although only 20 percent of our firms considered themselves to be in crisis situations, another 45 percent had either resigned themselves to continually decline or faced an uncomfortably fast expansion path which they felt they had to follow to survive. Thus, on the one hand, a wholesaler/dealer remarked gloomily that 'business is just good enough to keep going, but not bad enough to quit'. In contrast, the manager of an expanding manufacturing firm with thirty-four employees noted that the company was too big for one person to manage, but had to grow if it were to remain competitive. Another felt he would have to accept a take-over bid soon. Some companies, it seemed, were not totally happy about their future growth even when prospects look good.

Some firms consciously chose to remain small in the face of diverse conditions and factors. A successful liquor retailer did not wish to expand or branch because of family problems. A grocer explained his 'staying small' strategy in these terms: 'In the supermarket business, you either have to keep on growing or stay small to survive. I've adopted the strategy of remaining small – if you're small, you don't run the risk of over-reaching yourself and making major mistakes.' The decision to remain small was often associated with a risk averse attitude on the part of the entrepreneur. One retailer put it this way: 'I'd rather keep the doors open at an absolute minimum than face the possibility of bankruptcy and having to close down altogether; plus, I'm too embarrassed to furnish the redevelopment agency with my financial statement.' Sometimes a difference in attitude between owners and entrepreneurs could be detected. The manager of a manufacturing firm complained that his stockholders pressed him for dividends that constrained his ability to finance further expansion out of retained earnings. The survey's findings on the degree of innovativeness of small firms were mixed. While almost none of the present entrepreneurs had started their firms with 'a better idea', several of their fathers had. In each case, they had seen an unmet need that fitted the demand or supply side of their current business or occupation. For instance, the advertising firm had been started by a horticulturalist who began doing PR for agricultural clients, while a wood products firm moved vertically backwards into supplying itself with an input that was scarce. However, the set of entrepreneurs in the study had also demonstrated innovativeness in diversifying their operations, mainly through product innovation. Examples were the security products firm which developed new uses for its sewing/webbing process, and the construction firm which moved into oil-field pipe works. Some innovations in marketing strategies were reported – the book distributor contacted the entire array of international embassy trade delegations and stirred up more business than he could handle.

Very few of the innovations encountered represent improvements in an existing production process. One manufacturing firm did introduce a new process which lowered labor costs by 40 percent and increased employment through permitting a large expansion in the business. Another manufacturing firm invented a superior technical process for producing carpet tapes; this was also the only firm reporting an ongoing research and development effort. Our findings on innovation cast doubt on the applicability of product cycle models which stress product innovation as the province of new firms in youthful sectors and process innovation as the province of established firms and sectors (Abernathy and Utterback, 1978). While that model may characterize large firms in industrial sectors, it does not describe well the experience of our individual small firms.

The furniture dealer had an interesting 'recessionary theory' of innovation. He argued that it was much easier to start an innovative small business in the 'down' business period rather than at 'up' times. During boom times, companies were busy trying to serve existing accounts, could not meet existing orders for existing equipment or products, and had little spare capacity or time to entertain new ideas. But in recessionary times, companies had slack, were looking for new ideas which might increase sales or open up a new market, and were generally more receptive. He felt that now was as good a time as any to start a new small business (which he plans to do) – indeed, he felt it would be easier now than in 1949 (the year he began his now nearly defunct operation). His view is perhaps the small business counterpart to the theory of the bunching of epoch-creating innovations in the troughs of long waves (Hall, 1982; Mensch, 1979).

The net result of business conditions, competitive forces, up-turns and down-turns, structural changes, and management styles is a notable volatility in business history and profitability. Although responses to queries about profits were guarded and perhaps unreliable, the average small business reported a modest profit rate of between 5 and 10 percent. However, many experienced highly different profit levels from year to year, while others clearly were making profits quite in excess of the 5–10 percent range. These latter tended to register low net profits because a large percentage of their profits were immediately plowed back into the business and thus appeared as equipment purchases, expansions of inventory, etc. Several firms (for example, the recreational vehicle and furniture dealers) appeared to be passing through a product cycle from high profitability in earlier years to profit squeezes or crisis conditions now. For the very small firms and proprietorships, it should be mentioned that accurately assessing profit rates, as distinct from salaries, was difficult because of the large amount of time the entrepreneurs put into their business.

LOCATIONAL HISTORIES AND PRIORITIES

The evidence on locational factors influencing small firms generally confirmed those in published research. In only three cases had entrepreneurs moved into the Bay Area from outside because they spotted a market that could be served from this location. All the rest began or bought their businesses because they lived in the area. The reasons for this have much to do with the immediate circumstances of small business formation, specifically the origins of the entrepreneurs themselves. Most entrepreneurs interviewed did not fit popular images of the 'managerial type' capable of operating a business in any field or the 'inventor' starting out with a better idea to exploit. On the contrary, twenty-five of the entrepreneurs started their businesses after first-hand experience in that specific *product* line – eight from family or friendship experiences; thirteen from work experience as a manager, sales person, or engineer in that field; and four as workers (sometimes later rising to professional/managerial positions) for similar firms. Only two reported starting up in a completely new (to them) product line because they saw a market to exploit. Four of the entrepreneurs were 'pushed' into their business when their previous employer was on the verge of bankruptcy, divestiture, closing of a particular op∙ration, or elimination of their occupation (the real estate agency, a grocery, a manufacturer, and the plumbing supplier). Others were tired of working for someone else and saw a way to use their skills to branch out on their own. In general, then, most small businesses are not begun with an interregional locational choice; rather, they are spun off from already existing operations, large and small. A fair number of firms, however, had moved once or more *within* the Bay Area since their initial location. The factors conditioning these subsequent locational choices, in order of importance, seemed to be (1) market access, (2) adequate space at a reasonable price (typical of manufacturing, wholesaling, and durable goods dealers), and (3) access to transportation facilities, especially freeways (typical of those with area-wide or larger markets). Several entrepreneurs emphasized neighborhood amenities. The furniture dealer, for instance, liked a quiet business neighborhood as well as one with mixed residential/commercial land uses which he felt cut down his security problems. The book publisher had 'spiritual' needs, including the desire to escape pollution and to get as far as possible away from 'neighbors'. Only one firm, a larger manufacturing outfit, mentioned having to locate in proximity to workers' homes or transportation links.

Space needs varied tremendously depending upon the type of business. Desire for a larger, cheaper space was frequently cited as a reason for relocating within the Bay Area, although other firms were forced to move because their markets had shifted. Only seven of the firms owned in full or part the land and buildings they occupied. Those who did portrayed a variety of experiences with ownership. One decided to buy after having been evicted several times from previous quarters, at least twice due to redevelopment. However, others found that the illiquidity of being an owner posed long-term problems. One grocer, in a declining area, found that he was not only losing his equity in his building, but that he also could not afford to move because he was unlikely to be able to sell it at all. Furthermore, he could not even afford the operating costs of shutting down because, as the owner, he would have to pay liability insurance even if the land and buildings were unused.

Eighteen of the firms were leasees or renters of their space. These, too, had mixed experiences. Some would like to buy, but could not in an atmosphere of high interest rates and speculative land prices. Others complained of negligent landlords and of rising rents, although other firms had found good deals on rented or leased land that saved them money. The real estate agency, which sells space, felt that commercial rents were rising faster than the renters' ability to pay, which was bad business for them. Some leasing, of course, was purely formal – firms took advantage of tax breaks by setting up a sister corporation which owned the land and then leased it back.

In addition to these factors, the following hierarchy of needs with respect to capital, labor, and other factors of production also shape locational choices of small firms. Together they provide the basis for our recommendations on effective economic development tools in the final section.

RELATIVE SIGNIFICANCE OF INPUT COSTS AND AVAILABILITY

The business environment discussed in the first sections of the paper forms the greatest source of worries for small firms. Locational change, especially at the neighborhood level, often adversely affects a firm's market. But even when the demand for a firm's service or product appears stable over the near future, problems on the supply side may inhibit small firms' behavior. In our research, a clear hierarchy of supply-side problems emerged. Capital needs, both long- and short-term, topped the list. Labor requirements were only modestly problematic, while space needs (previously addressed), taxation, and government regulation were not generic issues of importance.

Start-up finance and capital

Access to capital has long been cited as a major problem for small business – a claim for which our survey evidence yielded strong support. Overwhelmingly, those interviewed pointed to the difficulties in procuring short- and long-term finance as the major internal factor hampering expansion and frustrating their ability to ride out cyclical, seasonal, and sectoral changes.

Capital needs can be characterized as either long-run or short-run. Long-term credit needs include start-up money (examined in the previous section), money to replace or expand fixed physical capital (machinery, equipment, new or expanded quarters), and money to expand inventories or make an extraordinary outlay that will substantially change the business. Most of our firms' experience with long-term borrowing was limited to the start-up effort.

Start-up (or buy-out) capital came from diverse sources. The most prominent, mentioned in at least twenty-three cases, was personal savings. These findings are similar to those reported by Storey, 1982. For many, this was the sole source of start-up capital. Five reported help from friends and parents as well. Only one had outside investors beyond personal connections. Five firms, mainly distributors, dealers, and retailers, cited supplier credit and advances of inventory. Four were extended credit by previous owners of the business who wanted to sell. One entrepreneur reported that customer credit helped him start up, one used a government loan (the health maintenance organization), and one (the book publisher/distributor) started with his own volunteer labor, living off his wife's paycheck. In some cases, entrepreneurs went to extraordinary lengths to start up their businesses. One entrepreneur took out a $1500 loan from a household finance company (at an annual interest rate of nearly 30 percent), one used his car and personal tools for collateral, another his furniture for the same purpose, while yet another took out a second mortgage on his home.

Six firms reported borrowing part of their start-up capital from banks. While this represents a minority (thus supporting small businesses' claims that banks will not generally finance business start-up) it does suggest that bank capital is possible. However, in almost all cases, the willingness of banks to lend at this stage depended upon either a relationship with a large well-known corporation (such as a supplier or previous owner) or the security of a type of collateral which the bank would find liquid if it were forced to foreclose, i.e. trucks, buildings, aircraft, real estate or recreational vehicles. Several of our firms expressed a willingness to borrow long-term but were very pessimistic about bank receptivity.

Many firms resorted to short-term borrowing to survive cash shortages that arose from supply rigidities, mismatches between payments and receipts schedules, and laggard accounts receivable. Nineteen of the firms – over two-thirds – stated that short-term cash needs were a serious problem. Two health service firms cited slow government reimbursement for patient services as a problem; a construction firm serving public sector clients also complained of unduly slow payment. Some sell to other small businesses who in turn have difficulty paying quickly. Both recession and inflation exacerbate cash flow problems – the first by increasing the magnitude of the problem and the second by raising the interest rates on short-term loans.

About one third of our firms stated that they had never borrowed a cent since opening. Some of them were adamantly and ideologically opposed to borrowing. 'The problem with loans is you gotta pay them back', was the succinct comment of a successful engine rebuilder with several branches. Others would borrow but could not find a reasonable source. For those that did borrow, interest costs ran quite high – one dealership reported that bank interest on his durables stocks ran to 40 percent of his costs. Of those using banks, a handful had succeeded in establishing a credit line, while others obtained intermittent loans for short- and long-term purposes. About four of the firms dealt with small banks. The six who borrowed from the big California banks felt that their creditworthiness was due to their track records (the case with branching efforts), their links with a large silent business partner (the security equipment firm), or the salability of their assets if their businesses should fail. Smaller banks were more apt to lend on long-standing personal relationships or on personal assets (a house, life insurance).

Not all borrowing took place with banks. At least nine firms received or currently used supplier credit. In some cases, firms who could not get supplier credit in the initial period have been able to get it subsequently through establishing a good track record. Supplier credit, however, generally came along with uncomfortably close control or scrutiny by the supplying corporation, and sometimes at a high interest cost. An example was the aviation dealer, who was forced to absorb overproduced inventory in the downswing and pay substantial interest on equipment not sold within ninety days. Food markets, on the other hand, had relatively amiable and less costly consignment arrangements.

The firms sampled exhibited a remarkable array of techniques for coping with short-term cash flow problems without resorting to suppliers or bank credit. Some adjusted input costs to changes in the business or product cycle or seasonality. The recreation vehicle dealer cut both inventory and labor in response to market saturation and the recession, while the packaging manufacturer drew down his inventory to accommodate insufficient demand. A small advertising firm used voluntary reductions in staff salaries to solve cash flow problems, with repayment plus bonuses promised in better times. Others adjusted their business payments and charge practices to handle cash flow difficulties. The florist cited his decision to accept credit cards as an antidote to accumulated accounts payable, despite the extra cost to himself. Two firms initiated interest charges on their accounts receivable. One firm, the furniture dealer, engaged in factoring accounts receivable, i.e. selling them to a collection agency. The plumbing supplier had decided routinely to charge other small businesses higher prices than his larger clients, because the former so frequently stretched out their payments when the business cycle depressed construction activity. Several firms reported stretching out their own payments to creditors, and in one case squeezing customers, as unpleasant but necessary ways of coping with cash flow crises.

For longer-term capital needs, firms sometimes resorted to structural changes in their form of operations. The Health Maintenance Organization (HMO) reported that many cash-strapped HMOs were merging with large insurance companies that were cash-heavy. However, retained earnings were the most common source of long-term capital for expansion or equipment replacement purposes. Many of the firms relied solely or largely upon retained earnings. It seems reasonable to conclude that the scarcity of longer-term finance and the dependence on retained earnings lead these small businesses to be undercapitalized on the whole.

Labor costs and related issues

Employment generation is one of the principal reasons for public interest in small businesses. The literature tends to assume that small businesses are characterized by low-wage, relatively unskilled jobs with high turnover rates, by hiring problems, by extensive use of part-time labor, by high degrees of occupational segregation by race, age, and sex, and by significant opposition to unionization. Our survey results contradicted many of these notions.

First, a large number of the jobs in the small businesses surveyed could be classified as skilled. These tended to be concentrated in the manufacturing, construction, and finance/insurance/real estate categories. Employers did not generally encounter difficulties in filling skilled job positions.

Second, three out of four were willing to do some on-the-job training. In a couple of cases, community college and technical school programs helped provide well-trained workers, and one or two had used government employment programs.

Third, a large majority of firms reported no turnover problem. Many of these had very low turnover rates, while others found no difficulty in replacing workers who had left. Of four reporting problems, three offered low-wage jobs with relatively undesirable working conditions. Confirming that neither labor shortages nor high turnover rates constituted a serious problem, almost all of the firms relied upon word-of-mouth recruitment. Drop-in job seekers provided a second source of new workers, and the State Employment Service yet another. Very few firms advertised jobs, and only one had used a private employment service.

The survey did uncover isolated cases of high turnover rates. One firm reported that its skilled workers, after learning and enhancing their skills on the job, frequently took advantage of opportunities to move on, often to competitors. A manufacturer prevented this problem from developing by deliberately limiting the skills his workers learned on the job. On the other hand, in some smaller firms (with less than ten employees), most of the workers mastered all of the business's operations and were thus highly valued by the management. One entrepreneur paid 'top dollar' to his workers to induce them to stay.

Several small firms found that hiring the *right* person for the job was often difficult. This did not necessarily reflect a lack of qualified candidates, but rather the combined effect of an inability to gauge workers' skills accurately beforehand and a reluctance to fire them subsequently. A mistake in hiring can be especially costly for a small firm. One employer hired a mechanic who had misrepresented his familiarity with diesel systems and had ruined a rig. ·

Fourth, a large number of firms offered full-time employment. Those who did offer part-time work did so because of the peculiar nature of their businesses, not in order to avoid paying benefits or meeting regulations. A decline in business, for instance, resulted in a firm (e.g. the furniture dealer) putting workers on a shorter work week, which was seen as more equitable (as well as more profitable) than firing. Alternatively, the advertising firm dealt with stagnation by eliminating artists' positions and contracting out to free-lancers. A manufacturing firm, a real estate agency, and a distributor all found it advantageous to use salespeople outside the firm's payroll to avoid carrying them in bad times. Seasonal ups and downs were occasionally met with by part-time help. And the time-specific services, such as restaurant meals, tended to employ part-time help. Nevertheless, most firms in the sample employed full-time workers with benefits as well as regular wage and salary commitments.

Fifth, findings on occupational segregation were mixed. Six entrepreneurs were proud of the level of racial and sexual integration of their workforces, although it was impossible to determine in all cases whether men and women, black, Chicanos, and whites were actually working in the same occupations with the same wage levels. Seven firms were clearly highly segregated both racially and sexually. Seven firms reported a racially integrated workforce that was either heavily male (six) or heavily female (one – a manufacturing firm with sewing operations). Only two firms exhibited gender integration but racial segregation. Summing up the categories, almost 50 percent of establishments were clearly racially segregated while nearly 70 percent were sexually segregated. Age segregation was difficult to detect – some firms hired in very limited age categories, while others reported workers of all age levels. Generalizations by sector and size were also difficult to make, although the smaller and/or more professional service-oriented firms seemed most apt to have integrated workforces. Examples were two groceries, a real estate firm, the securities investment company, and the Health Maintenance Organization.

Sixth, compensation rates were on the average substantially above the minimum wage. Wage levels among the firms were almost never reported to be below $5 per hour. Manufacturing firms tended to be on the high side; a wage of $5 or $6 an hour characterized entry levels, and only one paid just minimum wage. Relatively high wage levels do not appear to reflect shortages of prospective employees, but rather productivity, a relatively low labor content in several firms, a desire to keep out unions, and in a number of cases a desire to reward and keep able employees. Two firms planned to institute profit-sharing for their employees.

Unionization was not uncommon across the sample. Eight firms reported that some or all of their workers were unionized. These included all three groceries, a manufacturer, a construction firm, the Health Maintenance Organization, the plumbing dealer, and the restaurant. Most of these firms had no major quarrels with the union. Several used a union hiring hall for recruiting. One adamantly refused to hire from within the union but would permit his new employees to join; he was also the only employer reporting a past incidence of violence during an employees' strike. Fourteen of the firms had no union nor had experienced attempts to organize. Some went to great lengths to create personable relationships with their workers designed, at least in part, to preempt unionization drives; others paid higher-than-union-scale wages and benefits to achieve the same end. One firm 'got rid' of a union, while two others successfully blocked unionization efforts. The antagonism of certain employers towards unions arose from fear of a loss of control over the work process, rather than higher labor costs. Some entrepreneurs were frank about wanting to maintain strict control over their operations and strenuously objected to unions. In a couple of cases, managers argued that unionization would hamper the 'willingness to work.'

Public sector taxes and regulation

The debate on small business incentives often pinpoints government 'red tape' and excessive taxation as major impediments to survival. Our findings suggest that this problem has been overstated. These small businesses were less encumbered with taxes and (especially) regulation than is popularly suggested.

Of those firms responding directly, eight stated that tax levels were not particularly problematic for them, while only three complained about high taxes. Five others said that while tax levels were not a problem, the paperwork associated with filing was. Still others cited one particular tax or another as problematic – three cited the payroll tax, two the tax treatment of inventories, three the sales tax, one the proposed unitary tax, and one the unemployment insurance tax. For example, a retailer resented the sales tax being imposed on his purchases, but not on his sales. One owner lamented that he had not properly understood the proper procedures for charging sales tax. He had consistently not assessed it on a particular service item but then had to pay the back taxes on discovery. This entrepreneur argued that small businesses ought to receive better educational assistance from the government on their tax liabilities.

Small firms held divergent views of the favoritism of the tax system with respect to size. The restaurateur complained that small business did not get any tax breaks, while another complained that 'small businesses received no incentives to accumulate, nor to engage in research and development'. On the other hand, some firms seemed to think they received a relatively fair deal. The book distributor was thankful that imported books were not taxed. The high incidence of professional accountants preparing tax returns suggests that many do take investment tax credits and accelerated depreciation allowances. Partnerships and proprietorships take advantage of loopholes that permit the write-off of personal expenses as business costs.

Nor was there much consensus about excessive regulation. Only three firms claimed that regulation in general posed difficult problems for small businesses. At least six firms stated that regulation was not at all a problem. And, despite the publicity they have received, neither OSHA nor environmental regulations were mentioned by more than three firms each. Four firms (the advertiser, the paint store, the truck rental firm, and the tool sales distributor) said that their small size either helped them avoid adverse regulations or led to positive benefits from regulatory rules. Several firms found regulations cumbersome or costly but were willing to admit the real need for them, especially those which concern workers' health and safety or unemployment insurance.

Those firms complaining of regulatory ills generally had one enterprise-specific complaint. Two, for instance, found new energy regulations to be problematic. One, the small securities investment firm, had difficulty meeting all the SEC regulations. The construction firm resented minority contracting requirements. Two firms complained about heavy bonding requirements on large government contracts. One resented petroleum products anti-pollution measures; another disliked consumer product inspection. One grumbled about returnable bottles, another about liquor laws, another about unit pricing of groceries, while one grocer resented becoming an enforcement officer for the food stamp program. Many of these were not new regulations, and in most cases they were merely annoying, not crippling the firm.

In only a couple of cases did specific regulations pose serious problems. The real estate agency felt it would soon have to treat its agents as employees, resulting in large increases in social security and other payroll-related costs. In two cases, government-sponsored redevelopment activity had displaced businesses or severely crowded their markets. And, as we noted above, government tardiness in paying for health benefits posed large problems for two firms. But these examples are not, properly speaking, 'regulatory' ones.

Only one entrepreneur talked about engaging in illegal behavior in response to public sector problems. 'Government tax and regulation makes a person act illegally', he stated bluntly. He said that in his firm, it was possible to get around taxes by shortening inventory and underreporting sales (upon which sales tax has already been charged).

CONCLUSION

Our findings can be summarized briefly. First, the business environment of the sampled firms was quite turbulent, subject to seasonal, cyclical, spatial, and sectoral fluctuations of substantial magnitude. Particularly damaging were setbacks associated with rising costs, recession-related declines in sales, and market saturation at later stages of the product cycle. Many of the small businesses operated in markets dominated by large firms or depended on a few major suppliers or customers. These firms were quite vulnerable to changes in the business behavior of such larger firms. The strategies most often used by the small firms to deal with business environment and competitive problems included branching, product or service diversification, and product differentiation.

Second, sampled firms generally located in the Bay Area because their entrepreneurs lived there. A large number of them did change their business location within the area subsequently. Market access, adequate space, and access to transportation links appeared to be the most important determinants of present locations.

Third, finance capital posed a major problem for the firms, in both the long- and the short-run. Personal resources provided the sole or main source of initial capitalization. Only those firms with special connections or salable assets seemed able to secure bank credit. Cash flow problems were severe for over two-thirds of the businesses in the sample. Neither shortages of skilled and unskilled labor nor turnover rates posed major problems for the majority of the firms. Wage levels were not frequently cited as a deterrent to employment expansion. Unions were present in less than one third of the firms, and only firms in the manufacturing sector expressed strong opposition to unionization. Tax and regulatory problems posed fewer difficulties than business environment or capital access issues.

A survey such as this is useful for generating ideas and directions of work, rather than providing statistically valid results. Nonetheless, it does suggest some conclusions for research and policy. The theoretical grounding for small business research and policy needs to take into account the characteristics and the diversity that were evident even in this limited sample. Thus, the quasi-oligopolistic nature of small business competition, which is intimately associated with its localized character, indicates that an adequate theory of small business behavior should not assume perfect competition. In addition, the diversity in age of firm and business objectives suggests that we need to construct a more articulated model of the role of small business in urban structure. It seems to us that small businesses are clearly playing very different roles that range from community stabilization to providing vehicles for development and change.

For policy, a first conclusion is that we should be cautious in listening to advocates of specific proposals intended to enhance small business performance. The individual circumstances of small firms are extraordinarily diverse, and proposals favoring one group may well injure others. Nevertheless, the survey results do support some directions of policy. Capital constraints emerged here, as elsewhere, as the single most pressing problem. The forms of constraint, however, depend greatly on firms' individual situations. For example, lower taxes on reinvested profits will not help a firm caught in a cyclical squeeze on operating capital. The purposes of policy in this area need to be clearly defined. Similarly, the assumption that regulatory policy is oppressive to small business needs to be reexamined.

The observation that firms play very different roles in the urban fabric also suggests some possibilities for policy. Neighborhood conservation and revitalization programs should be designed to include a careful assessment of the needs and vulnerabilities of small businesses, and to provide explicit assistance if necessary. Economic development policy on a local level is likely to be far more effective if it recognizes the multiple roles of small business and attempts to respond to them.

Overall, we conclude that a stronger grounding in understanding of the real nature and diversity of small business is necessary for improved theory and policy alike. Characterization of this enormous and richly complex segment of the economy and society by a simple criterion of size is simply inadequate.

NOTES

1 The authors acknowledge the support of the California State Employment Development Department in this research. Patricia Carter, Amy Glasmeier, Jay Jones, Philip Shapira, and Douglas Svensson provided valuable research assistance. The views expressed in this paper are those of the authors alone.
2 Three basic job-size categories (1–5, 6–49, and 50–100) were employed. The basic aim of the study was to develop, in detail, a fairly small number of case studies rather than undertake a comprehensive and statistically verifiable analysis. However, an attempt was made to ensure that a reasonable cross section of firms (by sector and by size) were included in the study. The sample itself was constructed in two stages. First, a pre-test was designed to test the business information directory and the survey questionnaire. Second, a larger sample was subsequently drawn for use with a revised form. Since the business directory proved satisfactory and the survey revisions basically involved format changes rather than ones of substance, the results of the two samples were combined in the report. Both samples were stratified by selected Standard Industrial Classification (SIC) sectors and by business employment size categories. In the pre-test, the sectors covered were manufacturing, retail, and services, and the size categories were 1–5, 6–10, 11–25, and 26–50 employees. In the main sample, the construction, transportation, wholesale and financial, insurance and real estate sectors were added. SIC designation was controlled for only at the single-digit level and not for two-digit SIC industry groups. One exception to this procedure occurred in the manufacturing sector which was divided into durable and non-durable sections with an equal number of firms taken from each. Finally, the size categories in the main sample were restructured and expanded into three larger groups: 1–5, 6–49, 50–100.
3 *Contacts Influential* was used because the range of information it contains is superior to other business directories. Besides SIC designation and employ-

ment size, it was deemed important to know the key individuals in the firms in order to assure proper contact. *Contacts Influential* indicates for each establishment listed whether it is a headquarters, a branch, or a local independent. However, the employment figures presented in the directory for headquarters and branches refer to the establishment listed, rather than the entire firm. Thus, the auto repair firm was included in the pre-test with the expectation that it had 26 to 50 employees. In fact, the firm had a total employment of well over 100. Consequently, headquarters and branches were excluded from the main sample. Several of the local independents interviewed had branches of their own, but the total employment data were reasonably accurate as Table 8.2 suggests.

4 A number of sources proved helpful in the development of the case-study interview questionnaire, including the industry case-study methods used by the Political Economy of New England Project (investigators: Barry Bluestone and Bennett Harrison, Joint Centre for Urban Studies of MIT and Harvard, 1979); The Employment Needs Survey of the Oakland Private Industry Council (1980); questionnaires designed by Economic Research Associates (Oakland); and a study of small business (Kieschnick, 1979). Several pre-survey discussions were also held with small business entrepreneurs.

REFERENCES

Abernathy, William and Utterback, James. 'Patterns of Industrial Innovation. *Technology Review* **80**, June–July 1978; 40–7.

Armington, Catherine and Odle, Marjorie. 'Small Business – How Many Jobs?' *Brookings Review*, **1** (1982), 14–17.

Birch, David, *The Job Generation Process*. Cambridge, Mass.: MIT Program on Neighborhood and Regional Change, 1979.

Bluestone, Barry and Harrison, Bennett. *Capital and the Communities*. Washington, DC: The Progressive Alliance, 1980.

Fothergill, Steve and Gudgin, Graham. *The Job Generation Process in Britain*. London: Centre for Environmental Studies, Research Series 32, 1979.

Gordon, David. *The Working Poor: Towards a State Agenda*. Washington, DC: Council of State Planning Agencies, 1979.

Hall, Peter. 'Innovation: Key to Regional Growth.' *Transaction/Society* **19**, July-August 1982.

Ijiri, Yuri and Simon, Herbert A. *Skew Distribution and the Sizes of Firms*. New York: North-Holland, 1977.

Kieschnick, Michael. *Venture Capital and Urban Development*. Washington, DC: Council of State Planning Agencies, 1979.

Mensch, Gerhard. *Stalemate in Technology*. Cambridge, Mass.: Ballinger, 1979.

Storey, David J. *Entrepreneurship and the New Firm*. London: Croom Helm, 1982.

Teitz, Michael B., Glasmeier, Amy and Svensson, Douglas. *Small Business and Employment Growth in California*. Working Paper No. 348. Berkeley: Institute of Urban and Regional Development, University of California, 1981.

9

The implications for policy

D. J. STOREY

Chapter 1 emphasised the difficulties of obtaining a clear picture of the contribution of small firms to regional and local economic development. It was stressed that this was partly due to differences in definitions of the term 'small firm', partly due to differences in the levels of prosperity in the areas studied and partly due to differences in approach adopted by the researchers. Whilst in no way wishing to underestimate the extent of these differences, there also appears to be a high degree of consistency in certain research findings. In this chapter both differences and similarities are highlighted and a personal view is offered by the editor of the implications of these findings for public policy.

The impact which policies designed to create employment and wealth in small firms can have, within a time scale such as ten years, is discussed by virtually all authors. It is clear from these papers that the impact of such policies is very long term. For example Gould and Keeble show that in East Anglia 8,478 jobs were created in a decade. In Table 3.3 they compare employment in new firms in three counties of East Anglia (Norfolk, Suffolk and Cambridgeshire) with employment in new firms in Northern England (Durham, Cleveland and Tyne and Wear). They show that although on balance employment creation rates in new firms are higher in East Anglia than in Northern England, this is almost exclusively due to the 'Cambridge/Cleveland Effect'. The table shows that the contribution of new manufacuring firms to employment in Durham and Tyne and Wear was somewhat higher than in Norfolk and Suffolk, and it is the presence of Cleveland with its exceptionally few jobs created in new businesses in the Northern England group and the presence of Cambridge in the East Anglia Region that is the cause of these regional differences.

Even when the two Regions are compared, Table 3.2 of Gould and Keeble shows that at the end of a decade in East Anglia approximately 4.7% of manufacturing employment is in new firms, whereas in the North the comparable 'normalised' figure is 3.5%. In terms of employment this means that at the end of a decade, if the North had the same contribution from new businesses as East Anglia it would have created nearly 3,500 extra jobs. Placed in the context that a single decision by the British Steel Corporation involved the loss of 4,000 jobs on a single day in 1980 when it ceased operations in Consett, it is clear that differences in regional rates of new firm formation are insignificant as a short term explanation of differences in the economic performance of Regions.

Differences in new firm formation rates and the importance of new firms in employment creation are also discussed by O'Farrell and Crouchley. They demonstrate that in Ireland formation rates and the contribution of new firms to employment creation are significantly higher than for any British Region. However, it appears that the majority of new indigenous firms have been established in those sectors selectively sheltered from international competition whose growth is therefore constrained by regional and national demands.

Differences in industrial structure may also explain differences in new firm formation rates between Regions in the same country. For example, the higher overall formation rates in East Anglia, compared with the Northern Region of England, could merely reflect differences in industrial structure. The publication of VAT-based data[1] on structure of firms together with a regional and sectoral breakdown could, in principle, enable a formal shift-share analysis to be undertaken. Meanwhile, as Table 9.1 shows, the sectors in which new firms are found are similar in both Regions despite their different industrial structure. Differences in the rates of formation between the two Regions are, however, shown in virtually all sectors. Similar conclusions are reached by O'Farrell and Crouchley in their comparisons between Ireland and the East Midlands of England. They show that new firm formation rates in virtually all industrial sectors were significantly higher in Ireland than in the East Midlands, the only exceptions being hosiery and footwear – both of which are industries which, in the UK, are geographically concentrated in the East Midlands Region.

When comparing prosperous and less prosperous Regions, it might be imagined that the clearest distinctions would be in the presence or absence of so-called 'high technology' industries. The popular impression would be that the East Anglian Region, dominated by the Cam-

Table 9.1 *Sectoral distribution of new firms in East Anglia and Northern England (ranked by number of firms)*

	East Anglia			Northern England	
Rank	Industry	% of all new firms	Rank	Industry	% of all new firms
1.	Mechanical engineering	22.2	1.	Mechanical engineering	19.7
2.	Paper & printing	17.8	2.	Metal goods n.e.s.	14.1
3.	Timber & furniture	11.0	3.	Timber & furniture	13.2
4.	Metal goods n.e.s.	10.2	4.	Paper & printing	10.6
5.	Electrical engineering	8.1	5.	Clothing	7.4
6.	'Other' manufacturing	7.8	6.	Electrical engineering	6.4
7.	Instrument engineering	4.8	7.	'Other' manufacturing	4.4
	Total	81.9		Total	75.8

Table 9.2 *High technology new firms in East Anglia and Northern England*

MLH	Industry	East Anglia		North	
		No. of new firms	1981 employment	No. of new firms	1978 employment
271	General chemicals	2	22	8	98
272	Pharmaceutical chemicals	2	31	a	a
354	Scientific and industrial instruments and systems	26	253	12	350
364	Radio and electronic components	9	77	12	161
365	Broadcast receiving and sound reproducing equipment	12	201	a	a
366	Electronic computers	13	330	a ⎫	a ⎫
367	Radio, radar and electronic capital goods	8	151	15 ⎬	275 ⎬
483	Aerospace	0	0	⎭	⎭
	Total	72	1,065	56	991

a Less than 5 cases.

The implications for policy

bridge area, would have significantly higher rates of new firm formation and subsequent growth than the lagging Northern Region.
To test the hypothesis Table 9.2 takes the same MLHs classified by
Gould and Keeble as high technology and identifies the number of new
firms and their average employment in both East Anglia and the
Northern Region. It shows firstly that 72 out of 703 (10.2%) surviving
new firms were in high technology sectors in East Anglia compared
with 56 out of 774 (7.2%) in Northern England. However average
employment in these surviving firms in Northern England was actually higher at 17.7 workers than in East Anglian firms at 14.8.

In making these comparisons it will be recalled that the Northern
Region data were examined over a longer (thirteen years as opposed to
ten years) period so this may partly explain the higher employment
levels in Northern firms. On the other hand the fact that 'high
technology' firms have only been formed in significant numbers over
the past five years, and the fact that the Northern Region data do not
extend beyond 1978, may partly lower the formation rates in the
North.

Any comparison of the importance of 'high tech' firms in East Anglia
and Northern England must also recognise the importance of the spatial
clustering of such firms in and around Cambridge. Furthermore the
data presented refer only to *manufacturing* operations rather than to
non-manufacturing research and development, software and consultancy services, the provision of which *may* vary rather more widely
between the regions.[2]

Nevertheless these comparisons suggest that simple explanations of
differences in the contribution of new manufacturing firms to
economic development such as the impact of high technology firms or
other structural explanations are inadequate. It also suggests that
policies designed to improve the economic performance of regions
solely by the provision of incentives to create firms in certain sectors
would be to ignore more fundamental differences in economic
potential.

As Lloyd and Mason show, the differences between the new firms
formed in prosperous South Hampshire and less prosperous Mer-
seyside/Manchester are reflected in the personality, capacity and moti-
vations of the entrepreneur and other factors such as overall prosperity
of regional economy. Yet even these differences are *not* clear cut. As
Lloyd and Mason say:

The contrasts are not individually sharp but collectively they point to a
population of new firms which contains slightly older founders [in South

Hampshire], marginally more experience and creditworthiness and a small bias toward the use of more business skills in seeking the support of financial institutions and in having an awareness of marketing issues.

The general tone of this passage reflects accurately the nature of differences between regions in industrialised countries. It suggests that whilst there are differences between entrepreneurs between Regions these are often quite marginal and subtle, *but* that they impact across a broad spectrum of business. This suggests that a policy designed to encourage indigenous economic development cannot be restricted to a few sectors or a set of relatively narrow initiatives.

Instead, since the differences between entrepreneurs are marginal yet occur across a broad spectrum of industry, policy either has to be directed towards making marginal improvements in the marketing, purchasing, use of modern technology, encouragement for R & D etc. over both a large number and a broad range of companies *or* it has to be highly selective. In other words public policy has to be directed towards encouraging those few firms which have the ability to expand and by so doing both create jobs locally and also change local attitudes and aspirations.[3]

Public policies which are designed to impact upon a wide variety of small businesses are doomed to failure. The small businessman is, almost by definition, highly independent by nature, and the performance of the sector is diverse. Even in an area such as California, Markusen and Teitz comment on the diversity of performance of the sector and they emphasise the inadequacy of policies designed to assist all small firms:

... we should be cautious in listening to advocates of specific proposals intended to enhance small business performance. The individual circumstances of small firms are extraordinarily diverse, and proposals favoring one group may well injure others.

The diversity of the problems facing the small business, the independent nature of the entrepreneur, the fact that the major contribution to economic development in a locality is made by a handful of firms, all suggest that public policy cannot assist *all* small firms. These basic postulates are highlighted in Regions as widely divergent as California and Northern England. Public policy then has to be directed towards those firms whose improvement in performance, per £ of public assistance, has the maximum benefit for the economy.

This means discrimination. It means that those administering public policy have to avoid providing assistance either for firms with little prospects of growth or for those likely to cease trading. It also means that public funds should not be used to assist businesses which are selling the vast majority of their product locally since their growth will merely displace other existing local businesses. As businesses selling into non-local markets will depend upon the economy as a whole to provide market opportunities then policies to assist small firms will only work in a favourable macro-economic environment. Hence the objective of small business policy has to be to increase *net* employment locally; in short it requires a policy of avoiding the losers and picking the winners[4] within an expanding economy. 'Winners' in these terms will represent only a very small proportion of small firms.

Two familiar arguments are generally raised against such a strategy. The first is that it is extremely difficult to identify winners and that the public sector has a particularly poor track record in this regard. The second is that there is no point in assisting the winners since these businesses will succeed irrespective of public policy.

Whilst it is not possible to identify, with any degree of certainty, the characteristics of successful entrepreneurs the businesses which will transform an economy are clearly different, early in their lifespan, from the average business. Within five years of start-up those few businesses with the potential to transform the local economy are apparent. Furthermore a number of statistical models now exist which have demonstrated that the financial characteristics of failed businesses are significantly different from those of non-failed businesses. Such models could, in principle, be used at least to avoid making public investments in failing companies.[5]

The second argument that 'successful' businesses do not 'need' public assistance is also fallacious. Indeed a survey conducted by the editor (Storey (1985)) showed that new businesses which were most likely to encounter problems were, not surprisingly, those which were growing fastest. Frequently these problems were soluble by public policy. For example the rapidly growing firms required financial assistance, or help in recruiting labour or in obtaining new premises, etc. If public policy were able to ease constraints upon these rapidly growing firms the impact upon employment and economic activity locally could be considerable.

One area of public policy towards small business in which virtually all the authors in this volume suggest improvements is financial assistance. Despite this apparent unanimity it should be clear from an examination of these papers that finance for small business is far from a homogeneous commodity. This is clearly illustrated by Shaffer and

Pulver who show that whilst there are clear spatial variations in the availability of finance, the form of this availability or non-availability varied according to the type of finance. For example in some cases there were clear problems in obtaining equity capital, in other cases there were problems in obtaining loan (debt) capital and in other cases firms encountered difficulty in obtaining trade credit. Shaffer and Pulver show that there appears to be little regional variation in the ability of entrepreneurs to borrow equity, but in the provision of loan (debt) capital urban firms appeared more likely to be in receipt of both more and higher value loans than rural firms. Finally, in a most interesting table (7.13), Shaffer and Pulver show the sectoral differences in the provision of trade (supplier) credit, and how this form of credit may be a very important substitute for that available from the financial institutions.

The role of the financial institutions is also a cornerstone of the Oakey paper,[6] where he argues that the availability of financial resources is a key element in the innovative performance of high technology firms. Oakey shows that the Bay Area of San Francisco contains a large number of high growth 'high tech' firms, many financed by venture capitalists who themselves have established 'high tech' firms but who subsequently sold out in a takeover deal. The importance of the venture capitalist as a source of start-up capital is reflected in the fact that 30% of firms surveyed by Oakey in the Bay Area had begun with 'external' venture capital. This contrasts starkly with both Scotland and South East England which were also studied by Oakey.

These cases demonstrate that the financial infrastructure can be a major factor in providing an environment in which small firms can develop. Oakey, for example, believes that the combined presence of a locally based Bank of Scotland together with the willingness of the publicly funded Scottish Development Agency are major factors in the growth of 'high tech' firms in Scotland.

A responsive financial sector is then a pre-requisite for regional economic growth based upon indigenous potential and small businesses in particular. The form which that sector takes clearly varies from one Region or country to another. In Scotland the provision of equity capital has certainly been extended by the presence of the SDA. In major urban centres in England such as Birmingham and Leeds it can be made available by local government-funded Enterprise Boards, whereas in the Bay Area equity is almost exclusively privately provided. What remains equally clear is that in some areas the inadequate provision of finance could impede the birth and growth of small businesses.

It is, however, easy to criticise banks for being 'excessively' cautious in their lending policies, especially when the banks themselves were not interviewed.[7] Furthermore in many Regions the current financial infrastructure is a reflection rather than a cause of poor economic performance. Nevertheless if Regions are to generate growth internally a necessary, but not sufficient, condition is the presence of a financial infrastructure capable of responding quickly and flexibly to the variety of different demands placed upon it. The evidence presented by Oakey makes two separate points: first that the provision of this financial infrastructure does not necessarily have to come from the private sector, although it is clear in the case of the Scottish Development Agency that it has close links with Scottish private sector banks and financial institutions. Secondly, Oakey suggests that the prospects for the Scottish economy are significantly better than for most other less prosperous Regions and that this is to a major extent the result of a strong indigenous financial presence in the economy.

Three other points for public policy makers made by Markusen and Teitz merit underlining since other contributors also refer to these matters. The first concerns the nature of labour employed in small firms. Markusen and Teitz assert that the literature tends to assume small businesses are characterised by low-wage, relatively unskilled jobs with little turnover, whereas their own results suggested a large number of jobs in the small businesses which they surveyed were skilled. This, in fact, corresponds with the results of Lloyd and Mason who found that 36% of the workforce in small firms in South Hampshire, and even 33% on Merseyside, were skilled workers.

This finding, which corresponds with the editor's own work,[8] suggests that a policy designed to create employment through the stimulation of small business, particularly in areas of high unemployment, risks leading to a mismatch between those available for work and the requirements of employers. It may suggest the need for investment in an appropriate training programme, but volatility of the small business sector, together with the inability of small businessmen to forward plan, means that appropriate labour training programmes would be uniquely difficult to devise.

Second, the availability/non-availability of suitable premises is cited by Fothergill and Gudgin (1982) as a major explanation of the job losses which are occurring in British cities. The results provided by Lloyd and Mason for the Liverpool/Manchester areas do not suggest that a shortage of suitable premises 'explained' differences in the formation or growth rates of small businesses in either area. In Liverpool the public sector agencies were active in building small workshops, whilst in Manchester the closure of textile/clothing firms released considerable industrial workspace which was converted into small premises. The constraint was more likely to be binding in South Hampshire. On the other hand the Markusen and Teitz surveys did suggest that the performance, particularly of small firms in urban areas, was closely linked to the built environment of that area. Indeed, they are convinced that neighbourhood redevelopment schemes need to take explicit account of the impact which they have upon the small business population.

To offset such effects it is clear that small firm policies cannot operate in isolation from either government macro-economic policies or industrial policies. As far as the less prosperous Regions are concerned structural adjustment and a greater emphasis upon indigenous development can only take place under less restrictive monetary and fiscal policies than have been in operation in the UK since 1979. Even then progress will be slow and uneven. The lessons to be learnt from the United States are that the appropriate local financial infrastructure is a necessary condition for development. Beyond that public resources should not be dissipated in encouraging the creation of nebulous concepts such as 'an enterprise culture'; the unemployed cannot wait for two or three decades. Instead the full panoply of public assistance should be directed towards assisting that handful of firms in a locality that have the capacity to create significant numbers of jobs. In practice this means that some firms will wish to continue without any assistance, whereas the assistance package for others will vary considerably from firm to firm. It is, however, this diversity which is central to small firms and to public policy designed to assist such firms.

In summary despite the variety of areas studied and of methodological approaches there is a surprising degree of unanimity amongst the contributors. All recognise the increasing relative importance of small firms in creating wealth and providing employment, but concern is expressed that policies to promote small firm development will have differential spatial impacts.

NOTES

1 The sectoral distribution of new firms registered for VAT is given in Ganguly (1982a) whilst the regional distribution is given in Ganguly (1982b). Finally a lifespan analysis of firms is given in Ganguly (1983).
2 This point has been emphasised in correspondence with David Keeble, who believes the impact of 'high tech' firms is a major element in recent local economic development and performance. It remains an interesting, but as

yet untested, question whether software firms are more concentrated in locations such as Cambridge and the Thames Valley in England or Boston and Silicon Valley in the USA than the 'high tech' manufacturers.

3 This diversity of performance is stressed by Teitz *et al.* (1981) in their study of employment creation in California and by Mason (1984).

4 The editor's views on the importance of discrimination in implementing an effective small business policy are articulated at greater length in Storey (1983).

5 Watson (1984).

6 For a comprehensive review of the growth of high technology firms in the Bay Area of San Francisco and in Britain see Oakey (1984).

7 There have been a number of UK studies which have been critical of the lending practices of the commercial banks – many of them drawing upon the early work by Carrington and Edwards (1979). Nevertheless the banks have vigorously refuted much of the evidence presented – see Vittas and Brown (1982).

8 Storey (1982).

REFERENCES

Carrington, J. C. and Edwards, G. C. (1979), *Financing Industrial Investment*, Macmillan, London.

Fothergill, S. and Gudgin G. (1982), *Unequal Growth: Urban and Regional Employment Change in the UK*, Heinemann, London.

Ganguly, P. (1982a), 'Births and Deaths of Firms in the UK in 1980', *British Business*, 29 January, pp. 204–7.

Ganguly, P. (1982b), 'Regional Distribution of Births and Deaths in the UK in 1980', *British Business*, 23 April, pp. 648–50.

Ganguly, P. (1983), 'Lifespan Analysis of Business in the UK 1973–82', *British Business*, 12 August, pp. 838–45.

Mason, C. M. (1984), 'Small Business in the Recession: A Follow Up Study of New Manufacturing Firms in South Hampshire', Department of Geography Discussion Paper No. 25, University of Southampton.

Oakey, R. P. (1984), *High Technology Small Firms*, Frances Pinter, London.

Storey, D. J. (1982), *Entrepreneurship and the New Firm*, Croom Helm, London.

Storey, D. J. (1983), 'Regional Policy in a Recession', *National Westminster Bank Review*, November, pp. 39–47.

Storey, D. J. (1985), 'The Problems Facing New Firms', *Journal of Management Studies*, vol. 22, no. 3, July.

Teitz, M. B., Glasmeier, A. and Svensson, D. (1981), 'Small Business and Employment Growth in California', Institute of Urban and Regional Development, University of California, Berkeley, Working Paper No. 348.

Vittas, D. and Brown, R. (1982), *Bank Lending and Industrial Investment: A Response to Recent Criticisms*, Banking Information Service, London.

Watson, R. (1984), 'Local Government and the Business Sector: Local Authority Initiatives towards Small Firms', *Local Government Policy Making*, March, pp. 53–60.

Index

Abernathy, W., 205
ACE (Annual Census of Employment), 14–15
Allen, G. C., 45
Andrews, V. L., 166, 182
Armington, C., 2, 21, 166, 194
Atkin, T., 52

Bain, J. S., 116
bank finance, 158, 182, 210, 226–7
Bannock, G., 72
Bay Area, *see* California, San Francisco Bay Area
Beaumont, J., 64
Binks, M., 43, 52, 72
Birch, D. L., 1, 2, 15, 136, 166, 193, 194
Bluestone, B., 194
Bolton, J. E., 151
Boswell, J. C., 136
branches of small firms, in the Bay Area, 202–3
Brinkley, I., 43, 86
Britain
  small high technology firms, 135–65
  see also United Kingdom
British Business (1983), 72
Bullock, M., 149, 159, 162
Buswell, R. J., 75, 137

California, San Francisco Bay Area
  high technology firms, 138–9;
    employment growth in, 152–4;
    financial sources for, 151–2, 154–60,
    161; research and development, 144,
    145–9; small businesses, 195–217
Cameron, G. C., 44, 137
Cane, A., 69

capital stress, 187–9, 191
capital structure, of new small businesses:
  Wisconsin, 166–92
Cathcart, D. G., 59, 74, 76, 85, 116,
  117–18, 119, 124, 137
Cawdery, J., 78
Checkland, S. G., 74
Chinitz, B., 73
Churchill, B. C., 52, 115
Collins, L., 52
Cooper, A. C., 118, 128, 148
Council of Northeast Economic Action
  (CNEA), 166
Coyne, J., 43, 52, 72
credit denial, 185–7, 191
Cross, M., 4, 44, 56, 59, 60, 73, 76, 117,
  118, 119, 123, 125, 128, 151
Crouchley, R., 52, 103, 112, 220
Crum, R. E., 75
cyclical factors, in small businesses, 198–9

Daniels, B. H., 166–7, 168
Darnell, A., 116
Davidson, S., 168
debt capital, for new firms, 180–2
Deutermann, E. P., 138, 148
Dicken, P., 43, 44, 76, 78, 80, 85, 86, 88,
  95, 97, 102
Dunkelberg, W. C., 166

East Anglia
  new firm formation in, 43–71, 219–20;
  data and methodology, 45–6; high
  technology firms, 63–6, 68–9, 220,
  222–3; rate of, 72; spatial variations in,
  55–63, 67–9; temporal variations in,
  52–4

Eiseman, P. C., 166, 182
employers, *see* entrepreneurs
employment
    growth in high technology firms, 152–4
    prospects in new firms, 94–6
    in small businesses, 193–4, 211–13, 227
employment change, *see* job accounts
entrepreneurs
    attitudes to growth and innovation,
        203–6
    see also founders of new firms
equity capital for new firms, 179–80
Ewers, H. J., 135

Feller, I., 135
finance, sources of
    for high technology firms, 149–63
    for small businesses, 225–7; Bay Area,
        208–11
    see also capital structure
Firn, J. R., 43
Ford, P., 81
Fothergill, S., 2, 43, 44, 52, 53, 59, 61, 67,
        74, 75, 94, 136, 152, 194, 227
founders of new firms
    characteristics of, 75–6, 87–8, 116, 124
    see also entrepreneurs
Freeman, C., 135, 144

Ganguly, A., 43, 50, 152
Ganguly, P., 72, 77
Gibb, A., 77
Gibson, J. L., 148
GLC (Greater London Council), 2
Gordon, D., 194
Gould, P. R., 66, 219, 220, 223
government finance, 156–8
Greater Manchester, *see* Manchester
growth, entrepreneurial attitudes to,
        203–6
Gudgin, G., 2, 43, 44, 52, 53, 56, 59, 61,
        67, 73, 74, 75, 76, 94, 107, 115, 116,
        117, 121, 124, 128, 136, 152, 194, 227

Hall, P., 63, 206
Harris, C. S., 2
Harrison, B., 194
high technology firms
    in East Anglia, 63–6, 68–9
    innovation and regional growth, 135–65
    regional distribution of, 220, 222–3
Hoare, A. G., 77
Howick, C., 43

Ijiri, Y., 194

industrial composition
    of new manufacturing firms, 115–22; in
        East Anglia, 47–50, 220, 221; in the
        East Midlands, 113, 220; in Ireland,
        108, 110–12; in Northern England,
        220, 221
    of new small businesses: in the Bay
        Area, 195–7; in Wisconsin, 168–72
industrial in-migration, and new firm
    formation: East Anglia, 61–3
industrial sector, employment change by,
        17–19
inner city areas, manufacturing firms in,
        44
innovation
    entrepreneurial attitudes to, 203–6
    in small high technology firms, 135–65;
        and employment growth, 152–4;
        finance for, 149–52, 154–60; research
        and development, 140–9
investment finance, 154–6, 160
Ireland, Republic of
    new firm formation in, 52, 101–34, 220;
        comparison with UK, 112–15;
        international comparisons, 115;
        sectoral variations in, 108, 110–12,
        113, 129–30; spatial variations in,
        105–9, 125–8, 129; by town size and
        location, 103–5

job accounts, 11–12
    in the Northern Region, 12–40; data
        sources of, 12–15; by enterprise type,
        21, 22; by industry sector, 17–19; in
        new firms, 30–7, 39; since 1978, 37–8;
        by size of establishment, 19–20
job growth, in high technology firms,
        152–4
Johnson, P. S., 59, 74, 76, 85, 116, 117–18,
        119, 124, 137

Keeble, D. E., 62, 68, 83, 219, 220, 223
Kelly, T., 64
Key, T., 43
Killick, T., 85

Lever, W. F., 74, 77
Levi, P., 66
Lewis, E. W., 75, 137
Lirtzman, H., 166–7, 168
Little, A. D., 63, 78, 138
Lloyd, P. E., 43, 44, 59, 60, 67, 786, 78, 80,
        81, 85, 86, 88, 95, 97, 102, 223–4, 227
locational factors
    in Bay Area small businesses, 206–8

see also spatial variations
London Industry and Employment
  Group, 43

MacCracken, S. J., 1
McDermott, P. J., 77
Macey, R. D., 43
Macmillan Committee, 151
Manchester (Greater Manchester), new
  firms in, 79–81, 83, 84, 91, 92
Mansfield, E., 116, 135
manufacturing firms
  in East Anglia: formation of, 43–71
  in Greater Manchester, 80–1, 84
  in Ireland, 101–3
  in Merseyside, 78–9, 84
  in Northern England: employment
    change, 6–40
  in South Hampshire, 81–2, 84
  see also new firms
markets for small businesses, in the Bay
  Area, 200–2
Markusen, Ann R., 224, 227, 228
Marshall, J. N., 77
Mason, C. M., 44, 45, 46, 59, 60, 67, 81,
  85, 86, 95, 102, 128, 223–4, 227
Mensch, G., 206
Merseyside, new firms in, 78–9, 83, 84, 91,
  92
Mikesell, J., 168
Moseley, M. J., 55

National Economic and Social Council,
  130
new firms
  definition of, 45
  in East Anglia, 43–71, 219–20
  finance for, 149–63, 179–61
  high technology, 135–65; in East
    Anglia, 63–6, 68–9, 220, 222–3;
    employment growth in, 152–4
  in Ireland, 52, 101–34, 220
  in the Northern Region, 6–9, 219–20;
    formation of, 27–30; growth and
    death of, 30–7, 39; manufacturing, 21,
    23–7
  in South Hampshire, 81–5, 91, 93–4,
    96–7, 223–4
  in the United Kingdom: characteristics
    of founders/owners, 75–6, 87–8, 116,
    124; employment factors, 94–6;
    environmental factors, 92–4; financial
    context, 90–2; market context, 90–2
  see also manufacturing firms
Nicholson, B., 43, 85, 86

Northern England
  high technology firms in, 222–3
  manufacturing employment change,
    6–42
  new firm formation in, 6–9, 219–20

Oakey, R. P., 63, 75, 135, 136, 137, 138,
  142, 143, 145, 148, 226, 227
occupational structure of workforce and
  firm formation rates, 74–5; in East
  Anglia, 60–1, 63
Odle, M., 2, 21, 166, 194
O'Farrell, P. N., 52, 103, 112, 129, 220
O'Loughlin, B., 129

policies, public, and small firms, 1–2,
  219–29
Prakke, F., 149
product differentiation, 202, 203
product diversification, 202, 203
product life cycles
  in high technology firms, 140–2
  in small businesses, 199
Pulver, G. C., 226

Quince, T., 77

Rabey, G. F., 77, 78
Rainnie, A. F., 2
Rees, J., 135, 137
Reeve, D. E., 81
regional growth, in high technology
  firms, 135–65
regional variations
  in capital structure of new small
    businesses: Wisconsin, 166–92
  in new firm formation, 219–24
regulation, of small businesses: Bay Area,
  214–15
research and development
  in high technology firms, 140–9, 160;
    external, 147–9; internal, 143–7
Riley, R. C., 81, 135
Roberts, E. G., 136
Rothwell, R., 63, 135, 136
rural areas
  East Anglia: firm formation rates in,
    55–6, 61, 68, 124
  Wisconsin: capital structure of new
    firms, 172–91

Sant, M., 55
SBA (Small Business Administration) 166,
  167
Schumpeter, J. A., 116

Scotland
  high technology firms, 138–9, 160–1;
    employment growth in, 152–4;
    financial sources for, 151–2, 154–60,
    226; research and development, 144,
    145–9
  manufacturing firm formation in, 44
Scott, J. A., 166
Scott, M., 73
sectoral causes, of instability in small
    businesses, 199–200
sectoral variations, see industrial
    composition
Segal, N. S., 73
Senker, P., 142, 143
Shaffer, R. E., 225–6
Shapero, A., 166
Simon, H. A., 194
size of establishments
  employment change by, 19–20
  and firm formation rates, 59–60,
    74
Smith, I., 162
Smith, J-L., 81
South East England
  high technology firms, 138–9;
    employment growth in, 152–4;
    financial sources for, 151–2, 154–60;
    research and development, 144, 145–9
South Hampshire, new firms in, 81–5, 91,
    93–4, 96–7, 223–4
spatial causes, of instability in small
    businesses, 200
spatial variations
  in new firm formation, 123–8; in East
    Anglia, 55–63, 65–6, 67–9; in Ireland,
    105–9, 125–8, 129; in the United
    Kingdom, 43–5, 49–52, 72–100
  see also locational factors
start-up capital, 150–2, 208–11
Storey, D. J., 2, 4, 23, 27, 43, 60, 67, 73,
    76, 77, 116, 136–7, 151, 152, 209,
    225
strategies for small businesses, 202–3
supplier credit, for new firms, 182–5

Swales, J. K., 43

taxation, and small businesses: Bay Area,
    214
Teitz, M. B., 193–4, 224, 227, 228
Thomas, D., 63
Thwaites, A. T., 75, 135, 138, 143
Townsend, J., 136

United Kingdom
  new firm formation: comparison with
    Ireland, 112–15; spatial variations in,
    72–100
  see also Britain
United States
  new firm formation in, 52
  San Francisco Bay Area: high
    technology firms, 138–9, 144, 145–9,
    151–60, 161; small businesses,
    195–217
  Wisconsin: capital structure of new
    small businesses, 166–92
urban areas
  manufacturing firm formation in, 44, 55
  in Wisconsin: capital structure of new
    firms, 172–91
Utterback, J., 205

Vale, P., 52
venture capital, 159–60, 161, 226
Vernon, R., 135
Von Hippel, E., 136

Wedervang, F., 52, 115, 119
Wettmann, R. W., 135
White, L. J., 166
White, R. R., 66
Wilson Committee, 151
Wisconsin, new small businesses in,
    capital structure: regional variations
    in, 166–92
Wood, P. A., 44
Woods, L., 72

Zegveld, W., 63, 135, 136, 149

Lightning Source UK Ltd.
Milton Keynes UK
03 December 2009

147059UK00001B/19/P